EG Property Handbook

Geoff Parsons

2005

A division of Reed Business Information

Estates Gazette
1 Procter Street, London WC1V 6EU

ISBN 0 7282 0433 9

Typeset in Palatino 10/12 by Amy Boyle, Rochester, Kent
Printed by Bell & Bain Ltd., Glasgow

Contents

Part 1 Land and Ownership

Part 2 Property Markets

Part 3 Professional and Contractual Services

Part 4 Development of Settlements

Part 5 Development Process

Part 6 Property and Facilities Management

Part 7 Property and Finance

Part 8 Property Taxation

Part 9 Property and Government

Part 10 Settling Disputes

Part 11 Appendices

Part 12 Indexes

Dedication

To three special inspirers — Sophie Parsons and Carla Parsons
    For their unbounded curiosity and enthusiasms, and, particularly, their unstoppable use of sentences beginning: Why? Where? When? Which? What?
    And to Tomas Kai Newton

Preface

Talks with those who own or occupy residential or business property, either as owner occupiers, leaseholders or tenants or as investors, suggested that there is a need for a book about the property world. Awareness too of the needs of those who work with property on property-related matters, but without any formal training, has also suggested the need for a volume which describes how those in the property industry go about their work.

Establishing the limits of coverage of this volume has not been easy. Words such as property, construction, building, property management and facilities management have several meanings or nuances. In this volume the principal theme is realty or landed property, including buildings. The other types of property are touched upon but only in the context of real estate, eg stocks and shares are mainly referred to in Part 6 (on funding legal entities and property transactions).

Similarly, the use of the term property industry poses questions of the limitation of coverage in the sense that it may be thought to embrace construction, building or development, assuming that they are different. Also, the process of development may be used to embrace the construction process although both are industries in their own right. In this context the relationship of property and construction is treated in a two dimensional way. Part 4 covers the development process in four chapters. Part 3 shows both the property industry and construction industry underpinning a treatment of settlement development, town and community development, sustainability and building compliancy.

Finally, the seemingly new field of facilities management sits within the scope of the property industry in the guise of property management (in part), but some might argue that facilities management embraces the latter and much more. All of these are touched upon in their own right but the term property industry is conceived of as a synergy or gestalt for all: not one of them has a dominant or central position in this volume.

Finally, the work is intended, therefore, to be an introduction, even an aide-memoire, for those whose involvement in property requires them to understand or have insights into the ways in which property functions or processes are carried out by purely property professionals or others. The principal users of the book are likely to include:

- an owner or occupier of an existing building with a rent review
- a director or manager of a business facing the forthcoming revaluation
- an individual, eg the self-builder, wanting to start
- a representative of an organisation wanting to deal take a lease of property, perhaps for the first time
- the official wanting an outline of the government's involvement in the property market

- disputants — complainant and defendant — wondering how they will settle their difference
- a councillor facing committee papers on problems associated with the pension fund property portfolio or a planning application.

This is not to suggest that this volume provides every answer. What it will do is save the user time and hence, it is hoped, abortive expenditure. The need may be or is acute for the non-professional involved in property in that they may want to:

- understand the professionals' expressions or jargon
- identify their own problem in those terms
- know which facts, documents and opinions are relevant to their problem
- appreciate that in many instances a problem generates optional solutions
- sort out the general sequences and patterns of procedures followed by those in the property industry
- appreciate that actions (or inactions) have consequences.

Of course, not every one has a problem, the potential investor, dealer or other person may want to understand or get a feel for the property — what it is ... To some extent the book highlights some of the voluminous detail of the property industry with broad, general insights into property as a world of individual worlds within a country — the country is in the real world!

The book's 34 chapters (in 10 parts) are intended for the reader with an involvement — work, business, home or leisure — in property. However, property professionals and most students in property matters will, no doubt, find that the treatment of their specialist field is not deep enough. Nevertheless, the range and overall depth should suit any requirement for an all-embracing and readable text — particularly for the specialist who may require a handbook on hundreds of inter-related topics covering many aspects of the property industry — from which all may cherry-pick.

To end ... from a man of Kent ...

> All round and sound, Kentish Cherries!
> Who such cherries could see and not tempted be?

(From the Otley Collection)

Acknowledgements

The author of a book of this range of material has a heartfelt obligation to acknowledge those who have made it possible. Students, teachers and colleagues in both education and the property industry have continuously supplied me with answers to questions which have been raised over the years. Much of our discussions form the basis of each chapter.

Fellow members of committees and working parties at the Royal Institution of Chartered Surveyors in London and at the Kent Branch of the Institution provided me with valued insights to topics which were often originally beyond my knowledge.

American friends were more than generous with insights to not only planning law in the USA but also the British regime — one sparked a long term interest in marketing.

A special mention must be given of the late Vincent Cole FRICS who introduced me to writing and, similarly, the members of the panels for the two editions of *The Glossary of Property Terms*. We ranged, it seems now, every possible topic concerning the property industry, and more. The editorial digging refreshed long forgotten topics and introduced me to new areas — waste management, environmental land management, contamination and facilities management — to mention a few. A detailed structure for the latter was introduced to me by Bernard Williams in giving me the opportunity to work with him on his publications on Facilities Economics (for the UK, Europe and Australia).

Although I shall claim all of the errors (and omissions) as my own, I must especially, and gratefully, acknowledge the help of the very many who gave generously of their time for insights, comments, corrections and sources.

I am very appreciative of the reading or discussion on parts of the draft material by the following: Roger Bryan, Ray Craddock, Sheila Dent, Christian Schrant and Patricia Williams.

Also, I acknowledge with thanks the detailed discussion, comments and insights on particular chapters given to me by: Sarah Goodall, His Hon. Harvey Crush, a practising arbitrator and accredited mediator, Councillor Mark Fittock, Kent County Council, Professor Barry Redding, Albert Russell, and George Williams.

My appreciation of the patient efforts of those at the *Estates Gazette* in guiding my hand is recorded, particularly Audrey Andersson, Amy Boyle, Rebecca Chakaborty and Alison Richards.

Finally, my thanks to members of my family for all the day to day help, not least, to June for the almost always cheerful forbearance on delayed jobs in the garden and in the house!

Author's Note

This volume *EG Property Handbook* is in 12 Parts: 10 Parts comprising 34 chapters and two Parts of Appendices and Indices. The first 10 Parts are as follows:

1. *Land and ownership*: in this part the four chapters examine the nature of land, what property is owned in land, ie an estate or interests and rights in land.
2. *Property markets*: the four chapters look at owners, property marketing and the property markets: it also covers property information.
3. *Professional and contractual services*: the four chapters consider the nature of professional bodies, the work of their members, contractors and the nature of transactions.
4. *Development of settlements*: the two chapters here distinguish the development of settlements from the development of individual buildings, covering the wide areas of sustainability and compliance. They also deal with developmental aspects and management of settlements, and the regulations and controls used by the planning and other officials.
5. *Development process*: the four chapters seek cover the stages of the development process from the idea to the handover by the contractor.
6. *Property and facilities management*: three chapters distinguish the two managements in terms of commissioning, caring and insuring buildings.
7. *Property and finance*: three chapters on ownership and money go together in this examination of accounting, and appraisal.
8. *Property taxation*: five chapters in this part review the management of taxation and the national and local taxation.
9. *Property and government*: three chapters relate the nature and role of government at all levels to property ownership by looking at departments and agencies and their expenditure and funding.
10. *Settling disputes*: the final chapter describes the property industry's different ways of dealing with disputes and of handling a dispute.

In addition, it is hoped that all readers will find particular value in the chapters' illustrative boxes of which there are over 100. They cover the following:

* guides to professionals and what they do
* step by step guides to procedures
* profiles of organisations or other aspects of the property market

- case studies of marketing mixes — property
- information in concise but comprehensive text
- examples illustrating the text.

Parts 11 and 12 are appendices and indexes respectively — the readers' access mechanisms to the 34 chapters. They include a list of the boxes; lists of enactments; abbreviations; a few cases; and, three indexes — places, organisations and key words.

At the beginning of the book, the chapters' main sub-sections are shown under each chapter in the contents pages. Each chapter begins with an aim, ie the main reason for the chapter. The objectives follow as targets for the reader. As mentioned above, the text of each chapter is illustrated with boxes, providing a considerable, but concise, insight to the property industry

Generally, the use of a term follows the relevant meaning given in the second edition of *The Glossary of Property Terms* which is published by EG Books. (Alternatively, a term will sometimes be found on the publisher's web-site *www.egi.com*.)

Part 1

Land and Ownership

Land, Air and Sea

Aim

To explore the concepts of land, air and sea and their physical attributes and properties

Objectives

- to describe "land" (including water and air) as a legal concept
- to establish the extent of land at the surface, below ground and above the surface
- to describe the rules of law for determining the physical boundaries of land
- to establish the nature and physical properties of land, air, and fresh water
- to determine what is underground, eg minerals, other substances and hot phenomena
- to establish in relation to land the status of wild creatures, plants and other living things
- to describe the attributes of the sea and its plants and creatures

Introduction

For an isolated family in a pre-history economy, ie mainly hunting and cultivation for immediate living needs, land should be sufficient to provide space for all their needs, particularly for shelter and cultivation. If not, the family would move on to settle elsewhere. As the population grows land is less likely to satisfy their requirements and rights and obligations in respect of land are in the first instance desired and eventually realised by agreement or by force. In effect, the first steps to ownership in respect of land were being taken.

In common parlance, most people buy or rent a factory, a house or an office, and so on. In practice, they usually acquire a freehold estate or take or acquire a leasehold estate in land. Each estate will have unique and particular characteristics.

Thus, land ownership is the ownership of a complex bundle of rights and obligations. The bundle reflects the extremely complex society in the UK. However, land underpins all that is done. In the laws which apply to England and Wales land (not an estate, for the time being) has certain physical attributes which provide a foundation for the ownership of estates and other interests. (The laws applicable in Scotland are somewhat different.)

The air over land and water or under land also has physical characteristics which are dealt with below. Finally, the sea is somewhat different again, as will be seen later. In almost all instances of the use of the term land in this chapter refers to land not an estate in land.

The chapter begins with the idea that land is not owned. It deals with its natural characteristics and with those of air and water from the same perspective; but treating the sea somewhat differently. However, some points will be covered as if land is owned — but in a way which is obvious. Ownership of freeholds, leaseholds and certain rights concerning land are covered in chapters 2 to 4.

Land is incapable of ownership

Apart from one important exception land cannot be owned under the law covering England and Wales, ie English law. This is of little practical importance except when the owner of the freehold estate in a parcel of land dies without a will and there is no one to benefit under the rules of intestacy. In these circumstances the land "reverts" to Crown under escheat. (The "reversion" recognises the feudal origins of the freehold being granted to a baron by William the Conqueror or a later monarch.)

Crown's ownership of land

Thus, there is an exception to the general rule that land is not owned under English law; the Crown owns all land. Conceptually, this may be regarded as an example of what is known as "alloidal" tenure.

Alloidal land

An exploration (elsewhere) into the concept of alloidal ownership may be of interest. For the moment, it is sufficient to note that in some parts of the world land is owned, but usually by a large group of people. Indeed, it is conceivable that alloidal ownership existed in England in the Dark Ages when extended families occupied and cultivated small areas of land. Even today in the Orkney Islands of Scotland udal land tenure is said to be alloidal in nature. (It was introduced by the Vikings.)

Characteristics of land

Box 1.1 gives the characteristics of land. The principal characteristic is its support of life, ie its capacity to provide for shelter and for food production.

It could be argued that land satisfies human needs for shelter, food and water and, possibly, others.

Land's physical extent

What then is land? Conceptually, physical land extends physically from the centre of the earth to the heights of the heavens — somewhat like a misshapen cone, the sides of which pass through the boundaries of the land. In describing the concept of land in English law nothing has been (or should be) left out of Box 1.2, which shows the extent of land in English law. For emphasis, almost invariably, buildings are part of the land.

It may be noted that in some societies land has been conceptually separated from buildings, eg in

Box 1.1 Characteristics of physical land, air and fresh water

Tangible nature	• land and the buildings and structures put upon it have a tangible reality	• estates and interests which man attaches to the land and buildings are conceptual and intangible
	• also, physical air and water are tangible	• similarly, rights in water and air are conceptual and intangible
	• similarly, animals and plants are tangible	• also, rights in plants and animals are man-thought and conceptual and intangible
Permanent nature	• land is permanent and is virtually indestructible	• it may be "destroyed" by the action of the sea when permanently submerged
		• volcanic eruptions may cause land to be lost and beyond recovery
Capable of use	• use of land may be considered permanent	• contamination may result in permanent disuse
		• planning or other regulations may cause cessation of use
Fixed nature	• land is fixed in position permanently	• rivers or the sea may change one or more boundaries
		• climate change or earth plate movements may cause levels to slowly rise or sink
Sub-divisibility	• land is capable of spatial subdivision	• sub-division may be stopped by government action or natural force
Wealth store	• land may be used as a store of wealth	• land is capable of being destroyed, but rarely so
		• land is capable of being contaminated, so reducing or destroying value
Support for life	• land has the physical attributes to enable buildings and structures to be erected for shelter and other primary purposes	• a lack of warmth from the sun may make living conditions difficult or impossible
	• land, with air and water, enables the rearing of animals and growing of plants for food	• a lack of water may make living impossible
		• a lack of water or sunlight or both may make rearing and cultivation impossible
Inspiration	• landscapes, skies and waterscapes inspire artists and others	• lack of access may prevent or dull inspiration

parts of the Peoples' Republic of China, but this may be straying, in part, into the issues of the role of the state and of ownership. So here it should be emphasised that the list relates to the physical content of land.

Box 1.2 Extent of physical land, air and sea

Magna of the centre of the earth	• provides some practical use, having geothermal properties
Solid strata of rock and mountain formations	• may contain deposits of solid minerals, liquid minerals (such as oil) and trapped gas
Surface: land	• allows the growing of plants and rearing of animals • provides support for buildings and structures • allows transport and communications systems • provides nutrients for growing vegetation, natural or cultivated
sea	• provides support for floating installations or seaplanes • supports surface and submarine transportation • provides food
Underground and surface fresh waters	• provides water for consumption • provides water for growing crops • allows fishing • provides leisure opportunities • may become polluted • may be diverted naturally or artificially
Archaeological remains	• contains remains of buildings and structures, humans, animals and artefacts, including treasure
Buildings and structures	• provide shelter and warmth for living space • enables work to be carried out • provides storage accommodation • provides indoor and outdoor space for sport, culture, recreation and other leisure pursuits
Air and airspace (Above ground or enclosed within caves, buildings, mines and the like)	• sustains life that needs air • enables air travel by aeroplane, glider, rocket, balloon and the like

---

Boundary markers

Maps and plans

In one sense a boundary is merely a line on a map which delineates the extent of an estate or a field. It may thereby indicate ownership. Maps or plans are often not precise enough to show the exact position of a boundary. Hitherto, plans attached to most land registration documents should be regarded as indicative only but this will change for any of the plans which will, in due course, be created under the provisions for detailed plans in the Land Registration Act 2002.

Use of maps and plans

It is useful to regard a plan as sufficiently accurate to take measurements with some degree of confidence that they are accurate: "maps" should not be so regarded. However, although plans are usually to scale when they are first drawn, with the passage of time the paper may shrink, expand or become distorted. It is important, therefore, to check that they have a grid overlay or scale as a check (preferably in both directions on the sheet). Also, plans by convention, show the direction of north by an arrow, ie a north point. Generally, the standards are those used by the Ordnance Survey (OS) to produce plans.

OS plans have many practical uses in the property industry, eg pinpointing location in property transactions and in planning applications. Generally, purpose prepared plans will be required for construction and the laying out of grounds. Users should be aware that metric measurement should be used for almost all applications in the property industry, as a result of the Unit of Measurement Regulations 1995, SI, 1995, No 1804.

Physical boundaries

In this section boundary is described in terms of the physical attributes of borders to the area of land which is enclosed. The oldest enclosure boundary walls still in use are thought to be the pre-history boulder, stone and rock walls of Bronze Age times at Penwith in Cornwall. They are certainly evidence of occupation. It is interesting that they wander the landscape filling in gaps between large natural boulders; seemingly an intended permanent occupation — even perhaps a first-defining of alloidal ownership.

Most physical enclosures are variously designed freestanding fences or walls, eg a post and wire fence or a wall of stone; but it may be the wall of a building. In other instances the boundary may be indicated by a hedge or a line of trees. In some cases a ditch or a balk may be apparent. Where land goes up to water the boundary may be either the subject of an agreement or certain rules of thumb apply.

Finally, a common wall between two buildings in separate ownerships belongs to both parties. Works to a party wall may be governed by the procedures laid down in the Party Wall etc. Act 1996.

Certain rules of thumb apply in many instances where the documentary evidence silent or patently erroneous. Sometimes there are simply no documents which help. Also, it is possible that local customary rules may operate in some areas. Box 1.3 shows some of the rules for determining boundaries (assuming that there is no documentary or valid testamentary evidence to the contrary).

Disputed boundaries

In the event of a dispute over a boundary, professional help may be needed resolve the issue. In due course, the Land Registry will provide for those who need one the definitive line of a boundary for registration purposes.

Box 1.3 Rules to determine ownership of boundaries

Walls	• stones, rocks and even logs are likely to be taken to the edge of a plot or holding • outer edge of any footing or foundation is on the owner's land • a single wall supporting two buildings, eg terraced or semi-detached property, will be a party wall which is owned jointly
Fences	• fence stands on the land of the owner • rails, posts and other support features are on the inside of the fence (the good side faces out)
Trees and hedges	• planted inside the boundary • may be a poor boundary marker — tending to grow irregularly • often give rise to disputes with neighbours
Ditches	• digging took place on the owner's land along side and inside the boundary • spoil from the ditch was heaped on owner's land • any hedge was planted on top of the heaped spoil
River	• the boundary is along the centre of the river
Party wall	• a common wall between two buildings governed by the Party Wall etc. Act 1996, regarding works to it
Disputes about boundaries	• for a party wall — the Party Walls etc. Act 1996 • for a high hedge — the Anti-social Behaviour Act 2003 • for access for repairs etc. — Access to Neighbouring Land Act 1992 • for a boundary's line — possibly, the Land Registration Act 2002 (in due course)

Natural rights

As will be seen in chapter 4, natural rights attach to the ownership of a freehold estate in land. They have practical attributes which may be regarded as akin to physical land. The rights relate to:

- support
- drainage
- air
- water.

Certain other rights are not natural and the natural right of view does not exist! Nevertheless, views of the beautiful landscapes, seascapes and townscapes are protected in the following ways:

- designations — areas of outstanding natural beauty and other designations
- planning standards — height restrictions on proposed development, eg to protect views of St Paul's Cathedral
- infrastructure which is hidden — the buried highway at Stonehenge
- section 106 agreements (T&CP Act 1990)
- environmental impact assessments.

The nature of water

Natural water occurs in several forms either on the surface or underground, including:

- springs
- ponds and lakes
- streams and rivers.

(Dams, wells, desalination plants and other man-made sources or depositories of water are another story.)

The natural right to water attaching to a freehold is dealt with in Chapter 4. However, the owner of the freehold may, in general, collect and use rainwater falling on the land and to use water on, beside or under the land. Rights to the use of water may be subject to limitations, such as:

- a requirement to obtain a licence for abstraction from the Environment Agency
- the natural rights of others
- acquired rights of others, eg fishing rights sold to an angling club.

Minerals and hot phenomena

Minerals

In general minerals may be mined or otherwise extracted from land or dredged from water by various means. They are then treated to produce metals, fuels, chemicals, building materials or other useful products. Box 1.4 shows the main persons or bodies involved in giving the principal consent to an operator wishing to find and obtain the right to extract minerals. In some instances there may be more than one consenter.

In most instances there will be a requirement to obtain planning permission and sometimes an environmental impact assessment will be required.

Box 1.4 Identifying the main bodies giving consent to mineral extraction

Coal		• Coal Authority licences operators to extract coal
Gold		• Crown Commissioners give consent
Gravel	Onshore	• owner of the land authorises entry and works, subject to statutory consents • owner of severed rights to the minerals give consent, subject to statutory consents
	Offshore	• Crown Commissioners licence extraction, subject to other statutory consents
Oil and gas	Onshore and offshore	• DTI licences operators to explore drill and extract

Problems may arise from mines and other mineral workings, either from those being used or from disused workings or waste tips. These problems include:

- gases which escape
- lack of support to surface land due to collapsed workings (see natural rights, above)
- loss of surface water or underground water which percolate into workings
- contamination of the land or land which is nearby
- danger from unprotected shafts and derelict buildings.

In coal mining areas there is a statutory scheme for dealing with subsidence.

Hot phenomena

Hot phenomena is a coined expression to cover naturally occurring underground hot strata or heated waters. In the UK there is geothermal activity but it does not involve volcanic activity, merely geothermal strata which may give warm waters. Extraction of heat is usually induced by human activity. Such activity is treated as mineral extraction for the purposes of capital allowances.

Animals and other creatures

Wild animals

For the most part undomesticated animals and birds are wild and free to roam about the country unhindered. Similarly, fish are wild and free to roam waters in a natural manner. Nevertheless, many wild creatures are adapted to the presence of humans and may co-exist in a free but semi-tame relationship - they remain wild and not "owned" as such.

Domestic animals

Domesticated animals are another story and live for the most part in a long-term mutually dependent state with humans — they are not naturally wild and are owned unless cast into the wilderness.

Sporting rights

The earliest of human activities included fishing and the hunting of animals and birds and, of course, fishing. These activities are still very popular activities but not of necessity for most of us. Rights to indulge in these activities have a legal framework in "sporting rights" (see Chapter 4).

Concerns for animals

An outline of the position on concern with either wild or domestic animals is shown in Box 1.5.

> **Box 1.5** Ownership and wild creatures
>
> | Animals | • wild animals are not owned
• generally, wild animals are free to roam and must not be enclosed
• unless exempt hunting of foxes and other mammals with dogs is prohibited under the Hunt Act 2004
• certain animals are protected from hunting, eg badgers |
> | --- | --- |
> | Birds | • with one exception, wild birds are not owned
• the Crown owns swans
• most birds are protected
• for game birds there is an open season when they may be hunted |
> | Fish | • in a tidal estuary, fishing is freely available below the place where the tide ebbs and flows
• otherwise, fishing rights in a river or lake attach to the riparian land unless they have been sold or let by the owner |
> | Sporting rights | • angling, shooting and hunting rights may be attached to the land (see chapter 4)
• they may be held as a right, separately from the land
• a mere licence may be held to, say, fish |

Air

Properties of air

The proportions of gases which comprise air are generally stable. Above ground, air is freely available. It should sustain all forms of life, provided it is not polluted in the extreme, ie there is no undue change in the proportions of gases or other substances which are necessary for life.

Of course, in its natural state air is denser at the surface and therefore allows the sun to reach the surface of the earth in a varying intensity. Also, the movements of air (wind) carry the weather.

Airspace and its properties

If unrestricted, the airspace over land is available for working space to cultivate the surface or to construct buildings and structures which may be used for living, working or other activities. Obviously, the airspace is used to fly aircraft and to move vehicles or ships on the surface of the land or sea respectively.

Generally, the owner of land may use their airspace and restrict the use of it by others. However, Box 1.6 sets out the rights and limitations that the owner of the freehold has to the airspace above the land.

Box 1.6 Physical characteristics of and rights to air and airspace

Clean and safe air	• breathable air is a right
	• bye-laws may restrict bonfires
	• restrictive covenants may prevent bonfires, eg to protect a thatch roof
	• pollution by neighbours and others is actionable by authorities and perhaps the owner of the affected land
Air movement	• action of the sun on land and sea cause the movements of air and, hence, weather
Building rights	• building on land usually requires planning permission, building consent, and perhaps other consents
	• rights of tenants, neighbours and third parties may restrict building
Flying rights	• right to fly aircraft in the skies, ie above land is protected by the Civil Aviation Act 1965
	• manned balloons may land in an emergency, subject to making good damage
	• model aircraft, toy planes and balloons, gliders and kites require the owner's permission
	• byelaws may restrict or prohibit the flying of models
Right to flow of air	• there is no natural right to the uninterrupted flow of air (wind)
	• flow to an opening must not be obstructed when it has been obtained by grant or by prescription

---

Sea

The sea surrounding the British Isles has been a buttress for defence and a road for goods imported from and exported to countries throughout the world. It is also a play area for all kinds of sports and activities. Nevertheless, many aspects of the sea remain a mystery. It is not owned but the state claims jurisdiction over it within the limits agreed internationally. The sea's water, fish, seabed and other attributes are not of great import to most land-based folk, but those who want to use the sea for business or pleasure do so within a web of international treaties, European directives and UK enactments. This section draws some of the points together for those who may need some insight to the material. Box 1.7 sets out characteristics of the sea.

Territorial sea and the Continental Shelf

The UK government's jurisdiction over the sea generally extends to 12 nautical miles from baselines on land. Exceptions occur at certain places where the UK is too close to another country, eg France at the Straits of Dover. The jurisdiction is subject to navigation rights of those on lawful local or international business or pleasure. However, the seabed over which the UK has jurisdiction extends beyond the territorial sea to parts of the Continental Shelf. In this way the government has control, for

Box 1.7 Physical characteristics and other aspects of the sea

Ownership	• the sea is not owned
Jurisdiction	• the UK's jurisdiction extends to 12 nautical miles and to the Continental Shelf
Foreshore	• the Crown owns and manages most of the foreshore • the Duchies of Cornwall and Lancaster own part • others may own by grant or purchase
Seabed	• the Crown, DTI and local planning authorities have varied jurisdictions over the seabed's minerals • with existing or any forthcoming consents, minerals may be mined from the land or dredged at sea • North Sea oil and gas industry in decline • problems of decommissioning rigs, etc.
Quality and condition	• EU standards are used to measure the quality and condition of the sea • seawater is not potable but desalination plants make it drinkable • salt is obtainable from seawater • water utilities have a major responsibility to create treatment works for landward sewage (previously or presently discharged by pipes at sea)
Energy resources	• renewable energies forthcoming from both tidal and wave power
Coastal defence	• Environment Agency and others formulate polices for coastal defences and execute them
Authorisations	• for development, essentially the Department for Trade and Industry, Crown Commissioners and LPAs
Oil and gas	• industry is declining • gas storage under the sea is becoming important
Wind farms	• major seaward wind farm programmes of development are phased over the next 15 years
Fishing	• industry affected by EU competition policy
Defence training	• some near shore areas are affected by military training
Rescue services	• Maritime Coastguard Agency service coordinate sea and air rescue • Royal National Lifeboat Institute voluntarily maintain and operate rescue services around the coast of the UK
Warning services	• Trinity House maintain and operate coastal warning services
Charting	• the UK Hydrographic Office produces the Admiralty charts and other products for users of the sea here and elsewhere • it is a trading fund agency of the Ministry of Defence

instance, of the North Sea oil industry and the developing off shore wind farm industry. Also, waters within three miles of a baseline and coastal waters are controlled waters under the Water Resources Act 1991.

Ownership

Generally, the Crown owns the foreshore between mean high and mean low watermarks of ordinary spring tides. The exceptions are:

- ownership based on a grant from the Crown
- Crown ownership vested in the Duchy of Cornwall and the Duchy of Lancaster.

It follows that rights to extract minerals and to build on the Crown foreshore are subject to the Crown's permission. Similarly, development at the sea's edge may require Crown authorisation where it affects the foreshore. This will be in addition to any planning and requirements, eg an environmental impact assessment, of the local planning authority and other regulatory bodies.

Shoreline management plans

Planning of development of areas of land near the sea falls to the local planning authorities. However, there are a plethora of organisations with various responsibilities and perspectives, including:

- the National Trust which has been acquiring property adjacent the sea, eg under its Operation Neptune programme
- Trinity House owns and operates lighthouses and lightships on the coast and out of sea, respectively
- Maritime Coastguard Agency is responsible for emergency and rescue services
- the Environment Agency is concerned with the quality of coastal controlled waters and sea defences
- HM Customs and Excise are concerned with immigration control, smuggling of goods and human and drugs trafficking
- English Nature is concerned with nature conservation and protection, eg marine nature reserves and mudflats and saltmarshes created by coastal realignment and retreat (see chapter 13)
- Civil Aviation Authority need to safeguard air navigation control systems
- Department of Trade and Industry has licensing control of the oil and gas industry's exploration and extraction of the mineral
- the Crown which owns the foreshore and rights in gravel and other minerals.

Gravel and other mineral extraction

The regulation of extraction of seabed gravel and other minerals was the concern of the Crown Commissioners on behalf of the Crown. However, planning control has largely transferred to the local planning authorities. The Crown still has ownership rights in the foreshore and so on.

Rights to extract salt from the sea exist in some locations. Traditionally, from shallow pans (clay or concrete, etc the water is largely evaporated by air movements; boiling may be necessary in the final stages of the process.

When water from dredging is cleared of waste matter which goes to landfill, the waste is exempt from landfill tax. A Customs and Excise certificate must be in force at the time of removal.

Sea defences

The Environment Agency is the national body responsible for sea defences although owners of property adjacent to the sea will normally be responsible for obtaining planning permission if they want to erect sea walls and embankments at the sea boundary. A special capital allowance is available for new seawalls and so on, whereby the cost may be written off over 21 years. (Chapter 13 covers aspects of concerns about the coastal defence policies of the Environment Agency and other bodies.)

Oil and gas offshore

Various areas of the seabed around the British Isles contain substrata which bear oil and gas. The oil and gas areas are:

*   the North Sea off Scotland and the east coast of England
*   the English Channel off the Isle of Wight
*   parts of the Irish Sea off the coast of Wales and the seas off Morecombe Bay.

An oil and gas company wanting to discover oil or gas at sea must obtain a licence for exploration and development from the Secretary of State for Trade and Industry (see chapter 8).

Although the North Sea industry is now in steady decline a large number of oil and gas companies are operating there and there are known reserves and prospective developments in hand. As a result of the decline, the decommissioning of oil rigs and seabed installations is a major problem which the industry faces. At one time it was thought that the oil rigs might be submerged but they will need to be purified before submersion. Some will, it seems, be used in the newly developing industry of gas storage.

Gas storage

In some areas there is underground capacity for gas storage — as the indigenous industry declines the UK has become a net importer of natural gas. As a result it is forecast that substrata off shore (and on land) will be needed to store the gas as it is imported. This will also build up the nation's strategic reserves of gas. The needs are likely to result in a redistribution of gas-based facilities to concentrations around the mid-eastern coast of England and Milford Haven in South Wales — the latter to take liquefied natural gas (LNG) from Algeria.

Fishing

The fishing industry operates from a number of ports in competition from fleets from other countries. The industry is subject to European Commission quotas which amongst other causes have resulted in a decline in the UK's active fleet.

Marine nature reserves

Some areas are designated as marine nature reserves. For instance, the seas around Lundy Island are so designated by English Nature. It has a no-take zone to protect marine life to the east of the island

and also four zones with various management regimes for general use, recreational use and other protective categorisations.

Wind farms

The government's thrust to energy from renewable sources has resulted in a substantial growth of onshore and offshore wind farm development. A phased development of the off shore wind farms is in phase 2 of the programme controlled by the Department of Trade and Industry (see chapter 8).

The programmes are not without critics who are concerned with such matters as:

- interference with navigation rights at sea and in the air
- possible damage to fishing grounds and other life under the sea
- where the wind farms are on bird migration routes, possible harm to migratory flocks
- loss of sea views from coastal areas.

All proposals have required or will require the promoter to conduct an environmental impact assessment (EIA) and report it as part of the process of seeking approval.

Future

Technological advances in tidal and wave energies, together with the relatively mature windfarm industry, suggest changes, for instance:

- an undersea electricity grid and hub system
- many installations of tidal and wave devices to create energy
- an overall integration of planning and management of seabed conservation and other concerns about maritime developments.

Freehold Estates in Land

Aim

To explain the nature of the freehold and its variations

Objectives

- **to examine the origins of the freehold estate**
- **to describe the qualities or characteristics of the freehold**
- **to establish the nature of trusts**
- **to describe the nature of shared ownership**

Introduction

Historic basis for estates

Historic records show that land ownership was well established in the legal systems of Anglo-Saxon times. Writings of even earlier times confirm that property law was particularly strong in the Roman Empire. In the absence of written records before the Roman invasion it is not certain that the native Britons had a strong sense of an individual's owner-occupation of land or the ownership of an estate. It is conceivable that ownership was purely alloidal (see Chapter 1). From the time of the invasion, estates and interests were created under Roman law in those parts of the occupied by the Romans.

For the Dark Ages there is no evidence of surviving Roman settlements; although very early English poetry refers to ruins (probably Roman) but there was no indication of ownership of the land after Roman times. (Indeed, the poetry was, presumably, oral until it was recorded in writing.)

Again, one may speculate on the development of land tenure throughout the Dark Ages. Did ownership revert to local alloidal tenure by families or tribal groups? Did the advent of kingships over ever widening territories, eg Kent, result from power plays? If so, how was it achieved? Conceivably, alloidally held lands were progressively increased by several means, including:

- incorporation of unoccupied lands

- seizures of land held by defeated foes
- amalgamations of land through strategic marriages.

A king or queen who owned land would make gifts or grants of land in exchange for benefits, both to supporters and to those from whom support was sought. Forms of local feudal tenure probably developed in the Dark Ages but the introduction of Christianity resulted in written laws for several kingdoms in what was to become England under Alfred the Great. At that time the Danes had occupied much of the country and established many settlements — as many of today's place names testify.

However, the roots of today's land tenures lie in the conquest by William of Normandy. Under his rule a major reform of society took place in a relatively short time. Essential features of the feudal system under William included:

- a hierarchy of tenures, downwards from the barons immediately below the monarch
- specific services and goods being passed upwards from those holding lowest level tenure to the one above and so on
- a first national census of land and land-based resources, the Domesday Book
- development from that time of the common law, and of customary laws and rights.

Land and English law

Estates and interests are not the land

In Chapter 1 the emphasis was placed upon the physical nature of land which, as such, is not the subject-matter of ownership in English law: a person owns an estate or interest in a plot of land. These are conceptual objects, being utterly intangible in quality. Nevertheless, there is a common view that land is owned and for many practical purposes the distinction is not important.

Land and buildings as one

Again in Chapter 1, although land has the quality of supporting buildings and structures was mentioned, they are commonly thought of as being separate entities. However, it is an important maxim in English law that land and any buildings or structures on it are treated as one. Generally, therefore, any mention of land is a mention of the land and any buildings or structures erected on it. (In certain circumstances a particular building or structure may not be so regarded, and vice versa. This is a rare occurrence.)

Law of Scotland and elsewhere

The land law of Scotland is different than that of England and Wales — Scotland retained its own law after the Act of Union 1707. Tenures reflect more fully than elsewhere the ancient Scottish feudal laws; although progress towards land law reform is under way to change the system.

Generally, statutes on topics affecting land in Scotland are similar to those on the same topics for England and Wales — sometimes there are different statutes otherwise a statute may cover England, Northern Ireland, Scotland and Wales. Where the law on matter is dealt with in this volume, it has an

English law perspective — much of the practical content is broadly applicable to Scotland.

Freehold

The freehold estate or fee simple absolute in possession is the principal estate in English law. It has a number of inherent qualities which need to be appreciated for a meaningful understanding of it. They are explained in Box 2.1. However, other freeholds exist, eg life estate or estate tail.

Box 2.1 Qualities or characteristics of the freehold estate — fee simple absolute in possession

Derived from the Crown	• conceptually, land is derived from the Crown — originally by William I
Fee simple	• unlimited or unrestricted, eg does not depend on a life (as a life estate)
Indefinite period of ownership	• ownership lasts for ever, ie it is absolute
Capable of spatial sub-division	• estate may be divided into portions at the surface • any or all of the portions are capable of being transferred as freeholds
Ownership passes to heirs	• on death of the owner the estate passes to the heirs, either under the deceased will or under the rules of intestacy
Capable of alienation	• capable of being sold, gifted, exchanged and otherwise transferred
Sub-division into lesser estates and interests	• leaseholds may be created in the whole or in one or more parts of the physical property
Escheat (to the Crown)	• on death of the owner without a will or heirs (under intestacy), the estate reverts to the Crown
Possession	• is the capacity to control or use the property

Crown's status

Thus, a plot of land passed as a freehold under feudal tenure from the Crown at the time of the Conqueror (or from another monarch at a later time) and eventually came into the hands of the present owner. Ultimately, should the owner die without a will or a successor in intestacy, it will revert to the Crown under the principle of escheat. Escheat is, however, a rare occurrence.

Indefinite ownership

Freehold ownership is said to be capable of lasting forever but of course owners who are individuals do not. Companies, trusts and other corporate owners may last for a long time, indefinitely in fact and they may own freehold property throughout their life. In some circumstances, however, ownership comes to an end, not necessarily at the will of the owner but by some form of compulsion as indicated in Box 2.2.

Box 2.2 The forced loss of a freehold

Leasehold reform	• freeholds of certain houses may be bought (generally on favourable terms) by their lessees under the Leasehold Reform Act 1967, as amended • freeholds of blocks of flats may be bought by qualified lessees under the Commonhold and Leasehold Reform Act 2002
Compulsory purchase	• numerous Acts of Parliament enable local authorities and other bodies to seek the power to buy land and buildings compulsorily
Requisition in a national emergency	• in times war, the government has the power to requisition property for national defence purposes.
Escheat — passing on death	• if the owner dies without a valid will and there are no qualifying relatives under the rules of intestacy, the property reverts to the Crown by escheat.
Divorce	• court may order the sale of the property • court may order the conveyance to one of the spouses
Proceeds of crime	• court may make a confiscation order on conviction of a crime • court may make an order of sale of property representing the proceeds of crime
Bankruptcy	• An owner may lose property in the course of bankruptcy to pay off debts, etc

Natural rights

All freeholds have certain rights, so-called natural rights, ie air, drainage, support and water, which exist against others, particularly neighbours. They are dealt with in some detail in chapter 4. However, a right to a view does not exist unless special arrangements are made.

Enjoyment of a view

After the purchase of a property with an existing splendid view, it is not uncommon for the view to be marred by private or public development. While it may be a surprise, a freeholder of land has no natural right to enjoy a view from the property. Generally, when the view is lost there is no recourse to damages or compensation.

However, if the freeholder has the benefit of a restrictive covenant over land which is about to be developed, it may be possible to prevent the development. In cases where the developer of the burdened land has applied to the Lands Tribunal for a discharge or modification, the owner of the dominant tenement may argue against the change or, at least, seek compensation.

Apart from an existing restrictive covenant (or acquiring one) other means of protecting a view include:

- the purchase of land which is nearby, so as to gain control of its future
- the purchase of insurance cover against loss due to future developments.

Trusts involving land

From about the time of the crusades, trusts have generally come about because an individual or corporate body wishes to protect capital, so that the income from it or the capital itself may be used for the benefit of others. Family trusts and charitable foundations are typical examples. The idea of a trust is important and is, therefore, explored in some detail in this section.

The capital, eg, land and buildings, is conveyed by deed or under a will by an individual or an organisation to trustees who hold it, not for personal benefit *per se*, but for the purposes specified in the trust's founding documents. Box 2.3 summarises the creation of a trust and shows examples of the kinds of trust which may be created.

Formerly, several other types of freehold dating from the 1920s could be distinguished from the fee simple absolute, ie as trusts, such as:

- the fee tail
- the life interest
- the strict settlement
- the trust for sale.

In each case a person or persons were appointed as trustees — holding the property for the beneficiaries. Generally, as a result of the Trusts of Land and Appointment of Trustees Act 1996, persons with altruistic objects may still create trusts involving land but must create a trust of land. The strict settlement and other trusts can no longer be created and an existing trust for sale has been turned into a trust of land (strict settlements which were created before 1 January 1997 remain). Box 2.3 looks briefly at trusts and trustees.

Trusts and taxation

Although charitable trusts are not usually liable to taxation on income or capital gains, trusts, generally, can have complicated taxation contexts.

Sharing ownership

So far the position of a single owner has been dealt with in this chapter. However, it is possible for two or more persons to own the freehold in a property. Accordingly, issues arise when two or more persons intend to buy that do not arise when an individual buys. It would, therefore, be prudent to consider the types of or form of ownership in these circumstances. It is important before any commitment to buy together to establish the following:

Box 2.3 Points on the creation of a trust and examples of trusts which may be created

Wills and gifts (creation of a trust)	• a trust under a will comes into existence on the date of death whereas a gifted trust start as provided in the deed • selection of trustees and their successors must be considered with care — relatives, friends or professional or corporate trustees may be chosen • ensure clear objects are stated, the beneficiaries are specifically identified and the property or other assets are specifically stated or defined • understand the taxation implications of the chosen trust • considerer whether insurance is needed, eg for IHT • prudent to seek professional advice
Types of trust	**Accumulation and maintenance trust** • set up for children and grandchildren • limited to period up to 25th birthday of each child • consider taxation issues **Charitable trust** • a trust created to meet particular charitable objects • objects include alleviation of poverty, education, and public benefit • taxation beneficial • government policy on range of objects may change **Disabled person's trust** • a form of discretionary trust • for the benefit of a person who is disabled • tax benefits accrue for some beneficiaries (as defined in the Inheritance Tax Act 1984) **Discretionary trust** • each year trustees are free to give capital or income to beneficiaries as they see fit • trustees need not disburse to a particular beneficiary in a particular year • consider taxation issues **Employees' trust** • trust is created for the benefit of the employees of the grantor's business • grantor's business is conveyed to a trust • trustees run the business for the benefit of the employees **Fixed trusts** • beneficiaries receive fixed proportions of the annual disbursement **Leisure trusts** • some local authorities are setting up trusts to manage a leisure facility in their area • they are not-for-profit bodies **Co-ownership trust** • a joint tenancy is a trust (survivor takes all) • a tenancy in common is a trust (each tenant in common has saleable asset) **Protective Trust** • designed to protect the beneficiary from themselves, eg a person who reckless manages their financial affairs

	Service charge trust • a trust created to receive service charges from the tenants of, for instance, a shopping centre • ensure fund is in a separate account and for a specified building's requirements (to safeguard should the landlord become bankrupt) • monies used to pay for the services provided
Trustees	• the grantor needs to consider very carefully the role of the trustee and the knowledge, experience and other personal qualities of prospective trustees • the trustees will, in effect, become absolute owners of the property vested in them • they will be subject to the trust deed and the law, particularly the Trusts of Land and the Appointment of Trustees Act 1996 • the trustees actions are not only accountable in spirit to the grantor and the beneficiaries but to the law — the courts or the Charity Commissioners • the Trusts of Land and Appointment of Trustees Act 1996 sets out the duties and powers of the trustees • very briefly and generally, they may purchase a legal estate in land for investment, occupation by the beneficiary or any other reason • they must have regards to the rights of the beneficiary

- the shares of funds needed to buy the property
- the sharing of the costs of running the property and of disposal costs, including taxation
- the ownership position of each survivor as the others die
- the payment of any inheritance tax (IHT) when each dies.

Of course this may not be needed for spouses but for others it is prudent.

Shared ownership is as joint owners or as tenants in common or as commonholders; the first two kinds of ownership go back along way.

Buying property from co-owners

For a buyer of property from those in shared ownership, it is prudent to ensure that the monies are paid to at least two of the trustees of the co-ownership but this will normally be the solicitor's responsibility.

Joint ownership

In modern times spouses or partners tend to hold residential property as joint tenants. However, up to four persons may be joint owners. The essential quality is that on the death of one of the parties, the property passes to the one or more survivors. There is no process of transfer, ie contract and completion, other than a change in the Land Registry records. Of course, if there is insufficient cash or other realisable assets, any liability for IHT may result in a sale of the property. It may be necessary, therefore, to plan for the payment of IHT.

Tenants in common

A tenancy in common is similar but any one of the tenants may sell their share independently of the others. Also, on the death of a tenant in common, the share of the property does not pass to the survivor but is part of the deceased's estate that the personal representative deals with in accord with the deceased's will or the rules of intestacy.

Again, IHT may become due when a tenant in common dies. However, for many homeowners who are tenants in common the IHT burden may be reduced. When one dies their share in the property is liable to the tax, but for many the deceased's estate will come within the exemption limit and may be passed to others, eg a child, rather than their surviving spouse. (Of course other assets may bring the deceased's estate above the threshold.)

Taxation

Sales by joint owners where the property has been wholly or partly let, need to be evaluated for capital gains tax. Similarly, a sale of a portion between spouses or joint owners who are not spouses needs to be evaluated for stamp duty land tax — in the latter instance the buying joint owner may be liable for SDLT.

Commonhold tenure

In recent years leaseholders of flats have been able to buy together the freehold of their flats. The Commonhold and Leasehold Reform Act 2002 is one Act which enables the establishment of a new form of tenure — commonhold land. The Act also applies to shops and other types of property should the leaseholders wish to from a commonhold association.

The tenure is freehold and must be properly registered with the Land Registry. In effect they own it as a group and occupy the flats (or commonhold units, as they are known). The tenure gives them the "fee simple" and enables the enforceability of covenants for repairs and maintenance by the group.

The Commonhold and Leasehold Reform Act 2002 extends and develops the so-called right to manage (RTM) regime of earlier legislation. Thus, leaseholders in blocks of flats and other properties may create a company to take over the management of their property from the landlord. The provisions are quite multi-faceted and require complex accounting arrangements to be developed. The legislation became operational at the end of 2004 but it may take time for schemes to mature.

Leaseholds, Tenancies and Like Property

Aim

To explain the creation, change and termination of leases

Objectives

- to identify roles and activities in work concerning leases
- to explain the types of leases and their terms and conditions
- to distinguish leases, tenancies and licences
- to describe extensions, surrender and terminations

Introduction

The principal relationship between a landlord and a tenant is essentially one of property owner and occupying user respectively. Both parties have objectives which may change over time but at the beginning at least the landlord may simply want to ensure that the property is looked after properly and that the rent is received in full when it is due. The tenant will want either to make the property a home or be able to conduct business from it without undue interference from the landlord.

Written leases, contractual documents, are prepared by solicitors who must draft the lease with every eventuality in mind, virtually an impossible task. Apart from the detail of the terms and conditions which will be dealt with below, several important concerns have evolved over the years. Where the landlord and tenant have not been able or willing to deal with them, the government has attempted to solve them with legislation. The concerns have been:

- the level of rents for certain categories of lease or tenancy
- the tenant's security of tenure in the face of the landlord's demand for possession
- the landlord's day-to-day estate management in the interests of other tenants, business development and any estate property which is occupied by the landlord
- compensation, if any, which should be paid when the tenant gives possession at the end of the lease, for such items as the tenant's improvements to the property.

Apart from separate bodies of landlord and tenant legislation affecting agricultural property, business property and residential property, various private sector representative groups have attempted to reach consensus on the approach to particular concerns. These matters are dealt with below or in other chapters. Currently, some leaseholders of retail and other business property are expressing concern at the lack of regard by some landlords for the code for leasing such property; this may well become an area for legislative controls to be applied.

Roles and activities

Box 3.1 shows the principal roles and their activities involved in creating, managing and adjudicating on leases.

Box 3.1 Roles and activities in occupational lease transactions

Landlord	• prepares the property for letting
	• sets initial terms and conditions
	• repairs, maintains and insures, if required under the lease or a statute
	• receives rent and any service charges
	• may seek to resolve a dispute by a dispute resolution process
Tenant(s)	• pays rent
	• repairs and maintains and fulfils terms and conditions (as provided in the lease) subject to statute or agreed variations
	• may seek to resolve any dispute by a dispute resolution process
Solicitor	• drafts and executes the lease
	• registers leases of more than seven years
	• arranges payment of any stamp duty land tax
Letting agent	• promotes the property to let
	• receives applicants and shows the property
	• receives deposit
	• may prepare agreement for tenancies
Managing agent	• receives the rent and any premium
	• manages the carrying out of landlord's services
	• monitors the tenant's execution of the obligations under the lease
	• deals with any disputes between tenants
	• arranges cleaning of any common parts
Land Registry	• registers leases of seven years or more

Freeholds distinguished from leaseholds

Creation of a lease

A freehold is the highest estate in English law: a leasehold is inferior and is created:

- out of a freehold when the freeholder grants a lease for the first time; or
- when an existing leaseholder grants a sublease.

A freeholder may grant leases in different parts of the freehold as illustrated in the following examples.

Example 1: The freeholder of a multi-storey building grants a single lease for each floor on an internal repairing and insuring basis. The atrium, lift areas and corridors outside of each leased part become common parts shared by the occupiers.

Example 2: The freeholder obtains planning permission on a piece of land and builds an estate road and six factories. Leases are granted to industrialists for occupation on full repairing and insuring leases.

The basic principal features which distinguish a leasehold estate from the freehold are as follows:

- relationship of landlord and tenant is created under contract
- rent is payable to the landlord by the tenant
- terms and conditions in the contract define the relationship
- the lease must come to an end.

Types of lease, tenancies and licences

Several kinds of lease are granted by a freeholder or head lessees, the most common being the occupation lease for a property or part of a property.

Occupation lease and turnover lease

The occupying lessee will either reside there or occupy it for business purposes. Compared to ground leases, such leases are relatively short, being from eg three to 21 years. In recent years their duration has gradually reduced to not more than say, 10 years. A turnover lease is an occupation lease where the reference to turnover is to the kind of rent passing. The rent is usually a percentage of the turnover of the tenant's business with a minimum base rent. Turnover is usually subject to deductions.

Leases for the development of a parcel of land

When the owner of land grants a lease for development it is known as ground lease or building lease for say 99 years or 120 years — the latter term may allow for a second development after 60 years or so.

The ground lease usually provides for a specified period, commonly three years, during which the developer (ground lessee) is required to complete the building. If the work is not done the land should revert to the freeholder.

Licence to build and building agreement

Difficulties have arisen in obtaining possession under a ground lease, so an alternative approach has been formulated — the licence to build and agreement for a lease. Here the developer has a licence to build.

On completion of the works to the satisfaction of the owner of the land, a lease is granted to the developer usually on a full repairing and insuring (FRI) basis.

Box 3.2 summarises typical terms and conditions of:

- a licence to build (left hand side)
- an agreement for a lease (where the lease is usually FRI) (top of the right hand side).

Box 3.2 Terms etc of a licence to build, agreement for a lease and occupation lease

Licence to build	*Agreement for a Lease*
Duration of the building period	term of the lease
Attachment of: • plans • drawings • specifications	rent: • level • frequency of payments in the year • in advance or in arrears
Payments, if any	rent reviews • frequency • hypothetical usage • actual or hypothetical term, if any statutory obligations dispute resolution
	Occupational lease terms etc
Landlord's right to inspect	repairs and maintenance
Rates and taxes — liability	insurance
Licensee's default on works	service charge and renewals fund
Landlord's right to enter to complete the works	use or uses permitted
Insurances and/or performance bond	landlord's consents to tenant's • assignment • subletting • surrender
Lessee's right to redevelopment before the end of the ground lease	tenant's: • mandatory improvements • voluntary improvements
Reversion to the ground landlord	statutory obligations rates and taxes — liability of the tenant
Arbitration or other dispute resolution procedure	arbitration or other dispute resolution procedures

The head leaseholder may occupy the building but is more likely to let it. The lower part of the right hand side of the box shows briefly the terms and conditions of an occupation lease which might be used to sub-lease. (A lease for occupation does not, of course, require a licence to build, ie where the building to be leased already exists.)

Tenancies

Tenancies are somewhat less than leases but do establish a relationship of landlord and tenant. They are common in the agricultural and residential sectors, particularly the latter. In business situations they are, perhaps, less common. Traditionally, they cover periods of less than three years or are periodic, eg monthly. Agreements are often verbal and written agreements do not have to be drawn up by a solicitor (as would a lease under deed).

Despite their seeming informality, the relationships they portray — in terms of rights and obligations — have become woven with webs of lines of statutes. These have protected tenants in the main but do ensure the rights of landlords.

Licences

Licences are seemingly less formal and cover situations where the relationship of landlord and tenant is not intended but the use and occupation of land — even for a limited period — is paramount for the licensee. Typical examples include:

- self-storage use
- serviced accommodation for conferences and the like
- cinema visits
- use of sporting rights for limited periods.

Generally, accommodation shared with the owner will indicate a licence but this is not always the case. The importance of the distinction from an alleged licensee's perspective is that the statutory protections on the relationship of landlord and tenant are not available. In a given instance of dispute, legal advice needs to be sought.

Terms and conditions

In Box 3.2 terms and conditions of a typical occupation lease are summarised (see the second column which covers the FRI lease typical in a development situation). The landlord's solicitor normally prepares a draft lease which is then negotiated between the parties. Companies and others with a lot of investment properties usually have standard draft leases which are generally regarded as sacrosanct. However, several codes for lease terms and conditions have been devised for commercial leases.

Implied or invalid terms

In the absence of a full agreement between the parties, certain terms and conditions are implied by statute for certain types of lease or tenancy, particularly for residential or agricultural tenancies, eg the model clauses for agricultural tenancies. Similarly, statutes provide that certain terms and conditions are invalid for particular kinds of leases and tenancies.

Occasionally, a lease will contain an invalid clause, perhaps because a mistake was made. If the parties cannot agree to remedy the error, the lease usually provides for a method of resolving disputes, otherwise the parties may go to court.

Rental values and terms and conditions

The terms and conditions of a lease may have an impact on the rent payable when:

- the lease is granted
- on a review of the rent, or
- on renewal of the lease.

This section examines a few of the possible consequences of particular terms and conditions. There are many law cases which develop the principles to be applied so the topic is specialised and one in which some valuers and solicitors may be expected to offer advice based on considerable knowledge and experience of the field.

User and hypothetical user clause

A user clause, that is one which limits the actual use to one or very few uses, is likely to limit the market for the property and hence reduce the rent likely to be achieved. Similarly a hypothetical user clause, where the use for valuation purposes is not related to the actual use, will have an effect in line with whether it is restrictive or otherwise. A restrictive use will lead to a lower rent being agreed.

State of the property

Where works have been carried out by the tenant, it is important to consider the effect of any condition or state of the property clause. If the rent clause refers to the current state of the property (at the date of valuation), the works will usually be taken into account.

Duration of the lease

In a rising rental market it is likely that the longer lease will command a higher rent. However, if rent review periods are relatively short the rent will tend to be lower. In a lease, it is common for the rent review clause to have a notional duration assumed for the term.

Contractual and tenant's improvements

Contractual improvements tend to be included in the rent on review but wording usually provides that the tenant's voluntary improvements, if any, are a disregard.

On renewal of the lease under the Landlord and Tenant Act 1954 the tenant's voluntary improvements are disregarded unless 21 years or more have passed since they were done.

Statutory improvements

The tenant is often required to do statutory improvements and it is usual for the lease to regard them as a disregard on review. It may be noted that many occupiers of premises have duties of care under various statutes. Examples include:

- to ensure that adaptations are carried out under the Disability Discrimination Act 1995
- to ensure the protection of visiting workers and others against asbestos hazards, eg surveys and other works to establish the presence of asbestos.

Multi-let property

Some kinds of property are capable of being let to several tenants who occupy different parts of the building or complex. Examples include:

- purpose built flatted factories, blocks of offices or flats
- large shop premises divided into small shops or kiosks
- conversions into flats, offices and other uses
- out-of-town shopping centres.

Such clusters of occupiers are not without problems and these will be explained here.

Uses

For commercial and industrial property different uses in the same building or complex can cause difficulties for the some of the parties. The difficulties include:

- increase of insurance premiums (see below)
- conflict between retail tenants where one breaches a user restriction by selling goods or services reserved to another retailer
- the adoption by a tenant of a use which would not be authorised by the landlord because it would be in breach of a planning permission
- the adoption by a tenant of a use which would not be authorised by the landlord because it breaches use which benefits from a capital allowance.

Insurance rating of the premises

When a new tenant takes a lease care is needed to ensure that the use of the demised premises does not increase the overall insurance rating of the whole premises. Of course, the landlord could pass the increase to all the tenants though the service charge, but if this happens the long term effect to create a downwards pressure on the aggregate of rents for the property.

Shopping centre property

Letting of retail property in a shopping centre is a fully managed situation compared with lettings in a town centre where the landlord may only have one investment property. There are general principles of the management of retail premises which apply everywhere but are particularly appropriate in shopping centres. The tenant mix principles include:

- magnet tenants: usually large departmental stores which draw shoppers like iron filings to a magnet — they should be dispersed about the corners of the centre as far apart as possible
- cluster (comparison) tenants: left to their own devices such tenants will cluster together so that customers have maximum opportunities to compare, for instance, shoes — the shopping centre manager might prefer to separate them so as to churn the customers about the centre (food halls are the exception to the dispersal for reasons, perhaps, of such matters as cooking smells, deliveries and cleaning)
- complementary tenants: side-by-side these tenants will do more business than when separately located — their joint turnover will be greater (they can afford higher rents)
- interception tenants: some tenants are best located where they can intercept customers, say, as they leave the centre — possibly bulky items will be left to last, eg groceries as the customers go to the car park or the bus station.

Tenants' association

Where the landlord has many tenants it may be easier to manage them by encouraging them to form an association to act on their behalf. The benefits are likely to include:

- both parties will reduce the number of interactions between them
- they may save management's time (and fees)
- if not already, all the tenants could be represented professionally
- co-operation on centre promotion and other cooperative endeavours could be undertaken.

Disputes between tenants

Unless disputing tenants can settle their differences the landlord is likely to be drawn into the dispute, perhaps as arbiter. This will be particularly true where the dispute centres on a breach of a user clause, as explained above.

Rents of commercial and industrial property

New rents

Rents for new leases of commercial property are generally settled by private treaty negotiations between the parties or their professional valuation or property management representatives. Some of these rents are determined by tender.

Rents on review or renewal

The settlement of a new rent at the time of a rent review or near the end of a current lease of commercial property on its renewal will again be negotiated. In the event of disagreement as to the correct basis or the level of rent, a dispute resolution procedure will be triggered unless the tenant decides to give up the lease, either by using a break clause (if any) or by agreeing a break with the landlord.

Interim rents

Where protracted negotiations are taking place at the end of a lease an interim rent may be settled until the appropriate rent is settled.

Effect on rent of improvements

The rent review clause usually provides that any voluntary tenant's improvements be left out of the lease provided that the landlord agreed to them before they were undertaken. On a renewal of the lease, the Landlord and Tenant Act 1954 provides that they are ignored for any renewal within 21 years of the date of the improvement — the tenant must reckon to amortise expenditure on voluntary improvements within a particular period, ie the 21 years plus (possibly) a further period depending on the circumstances. For instance, at best renewals are allowed every 14 years (under the Act) so the period could be indefinite provided the landlord did want or could not obtain possession. Of course, where the landlord can obtain possession, the tenant may not enjoy the full 21 years.

Turnover or percentage rents

As indicated above, retailers in some shopping centres agree a basis for the rent to be paid which is a percentage of the annual turnover of the business. A base rent minimum is normally inserted, say, 80% of the open rental value. Turnover is carefully defined in the lease; for instance, deductions from gross turnover would include value added tax and tobacco duty (for a tobacconist). Rents at motorway service stations have usually been leased on a turnover rents basis.

Surrogate rents

The rents of some properties are difficult to value because there is a lack of comparable evidence. This may happen when a niche property sector is forming. Surrogate rents are adopted; for instance, retail warehouses were valued originally at a multiplier of the rent of ordinary warehouses.

Repairs and maintenance

The lease will lay down the duties of the parties for repairs and maintenance. Leases tend to be:

- full repairing and insuring (FRI) — where a single building is leased the tenant does everything
- internal repairing — the tenant is responsible only for the interior of the leased accommodation
- fully inclusive — the landlord does everything.

Landlord's expenses are frequently recovered by the payment of a regular service charge by the tenant.

Dilapidations

At the end of the lease, there are circumstances when the landlord will want possession of the property and serve notice on the tenant before it ends. If the tenant intends to leave, a liability for repairs, dilapidations, may arise, requiring the tenant:

- to make good the property, or
- make a payment to the landlord.

The standard of work is laid down in the lease but the courts have interpreted the meaning of such clauses from time to time. It may be appropriate when differences arise for the parties to seek professional advice. In the event of the tenant's failure to do the work the landlord may seek redress.

Service charges and renewals fund

Where a building is multi-let the leases usually provide for the landlord's expenditure on services to be recovered by a service charge being paid by each tenant. Similarly, the landlord will endeavour to recover the cost of any renewals, eg the replacement of a lift. Several approaches are used by landlords to recover the money spent on the property on behalf of the tenants. In chapter 23, Box 23.1 analyses the various approaches.

Taxation

An issue for the landlord and the tenants is taxation. Any payment of service charge or renewals fund not spent in the year in which it was received will usually be subjected to income tax or corporation tax. Any accumulation fund will be taxed as each payment is received, and the interest earned will also be taxed. Various approaches have been devised to address the problem so as to avoid tax.

Insurances

For a FRI lease the tenant is responsible for the insurance and should submit evidence every year that the annual premium has been paid. In other cases the landlord insures and the interest of the one or more leaseholders should be noted on the policy.

Application of monies

Care should be taken to ensure that the money proceeds of any claim are put to reinstatement of the damaged or destroyed building. The tenant will not normally expect to pay rent on vacated property while repair is being undertaken. A loss of rent or cessation clause needs to be included in the lease for this purpose. Insurance cover may be obtainable for the landlord's loss of rent during the period of the works to reinstate the premises.

Improvements

Improvements to property are either done by the landlord before it is to be put on the market, eg complying with statutory requirements, or when a prospective tenant wants alterations done before moving into occupation.

Agreed improvements

The improvements done by the tenant are known as agreed improvements or contractual improvements and become a condition of the lease (they are not tenant's voluntary improvements). They are executed and usually paid for by the tenant after taking the lease. The landlord will usually recognise the benefit to the property (and the landlord's reversion after the lease has come to an end) and may, therefore, grant a rent free period for say six months in recognition, or let the property at a lower rent than otherwise until, say, the first rent review.

Tenant's voluntary improvements

When the tenant wants to carry out improvements during the term the landlord's permission is sought and the work may go ahead. They are known as voluntary improvements and will usually not be taken into account when the rent is being reviewed. (The effect on rent on review and on renewal is more fully dealt with above.)

Landlord's access

The lease will usually provide for the landlord to have access for inspection of the state of the premises on giving reasonable notice to the tenant.

Alienation

Alienation involves the lease being disposed of by the tenant, essentially, with the landlord's consent. A number of different transactions may be agreed by the parties to a lease, ie the landlord's permission will normally be required. A permission may be implied or imposed in law; it frequently depends upon the wording of the lease. However, where the existing lease has value in the open market, the tenant may enter into one of several transactions, thereby realising it:

- as profit rent or a premium (or both), by subletting the whole or part of the leased property, ie the demised premises, as it is known
- as a capital sum, by selling (assigning) the lease
- as a capital sum, by surrender for payment by the landlord
- as latent profit rental value, by surrendering the existing (short) lease and taking a new longer lease (the terms may include monetary consideration).

Value and taxation implications

Each transaction has both value and taxation implications for both parties and the landlord may be concerned about the prospective new tenant (as sub-tenant or assignee) *vis-à-vis* any other tenants and the property letting. Thus, both parties should be aware of the taxation implications for themselves and for each other (as negotiating points) and ascertain for themselves the best net-of-tax outcome of the different transactions.

Security of tenure

For business property, generally, at the end of a lease the tenant will often wish to renew the lease and the landlord will frequently be willing to allow this to happen. Due to statutory interventions in the past, security of tenure differs in the various property sectors, ie business, agriculture and residential, and even within a sector differences of treatment occur. Nevertheless, there is a commonality of principles in that:

- there is a presumption in favour of the tenant retaining possession or, possibly being offered suitable alternative accommodation
- if the tenant leaves there may be a requirement to undertake repairs to a certain level
- there is a strict procedure which must be followed as to dates and information which may be needed
- the landlord should have possession on certain conditions being observed
- the tenant should have compensation in certain circumstances.

However, there is a somewhat uneven treatment in the different sectors, ie business, residential and agricultural. The detail is intricate; for instance, a notice which was served on the former agent of the former landlord was held not to be a valid notice.

Leasehold reform

The Leasehold Reform Act 1967 as amended and other legislation, eg the Commonhold and Leasehold Reform Act 2002, gives lessees of houses and flats respectively, additional security of tenure. Those in houses, provided they are qualified, may buy the freehold or extend their lease by 50 years (subject to certain rights afforded to the landlord). Similarly, those in flats may (as a group), become commonhold owners of units (see chapter 2).

Rights of Others 4

Aim

To establish the kind of rights which affect property

Objectives

- **to identify and explain how rights attached to land originate**
- **to describe the different kinds of rights attached to land**
- **to show how to defend, modify or extinguish rights attached to land**
- **to describe other rights affecting land but not attached to it**

Introduction

The heart of this chapter lies in the idea of the dominant tenement and the servient tenement, but covers rights in addition to rights between tenements. Although the title to this chapter refers to the rights of others, it will cover every landowner. Thus, every landowner owns land which is burdened by one or more rights attached to another person's land and at the same time enjoys rights which burden another's property. A piece of land so burdened is known in law as the servient tenement and the land with the benefit of a right is called the dominant tenement.

Unless otherwise stated, all rights are enjoyed by someone but they are, in law, usually attached to the land itself and cannot therefore be transferred separately from the land. There are several exceptions to this principle, eg a lordship of the manor.

Generally, most of the rights dealt with in this chapter are landed rights but some, the exceptions, either belong to a person as a matter of law or may be exercised by an individual in the course of a statutory duty, eg the police have powers of entry.

This chapter deals with the creation, maintenance, modification and extinguishment of rights.

Roles and activities

Box 4.1 shows those involved in holding land either enjoying rights or suffering rights. Where officials and others have a right of entry to land this is shown in Box 4.2.

Box 4.1 Persons involved in rights affecting land

Owner of the dominant tenement	• enjoys the right
Owner of the servient tenement	• suffers the right
	• may seek to modify or discharge a restrictive covenant
Solicitor	• on sale or purchase of property, will advise on existence of rights
	• deal with the creation, modification or discharge of a right
	• advise on protection of a right
Valuer	• advise on the price of acquisition of a right
	• advise on compensation on compulsory purchase of land affected by a right (to owner of the right)
	• s 84 discharge or modification of easement (Law of Property Act 1925)
	• compensation for effects of physical factors (Part I of the Land Compensation Act 1973)
Building surveyor	• advise on party wall matters
Insurer	• provide cover to developer where there is an old restrictive covenant
Person seeking adverse possession	• may claim title to the property
Person claiming prescriptive rights	• may be able to gain right to light or other right
Trespasser	• occupier's liability holds for a trespasser
Squatter	• may claim title to the property after due period

Rights and obligations which burden land

Apart from a lease, tenancy or licence, a property may enjoy rights or be burdened by one of the many other rights of a neighbour or other person. For instance, the owner of one property may have the right to walk a defined path across a neighbour's property. Where they exist, these rights are enjoyed by certain categories of person against the free use of a property by its owner. Each has arisen in the past as a result of:

- the common law, eg natural rights
- an ancient custom
- a grant by the owner or some other action, eg the grant of a restrictive covenant
- adverse possession
- prescription under the law
- a statute, eg right of access to land afforded to an official.

Natural rights

All freeholders enjoy natural rights against the land of neighbours, ie the action of adjoining landowners. They are a fundamental part of the freehold and do not have to be acquired separately. Natural rights include air, drainage, support and water.

Air

The landowner's right is to have any flow of air to his property in an unpolluted condition. There is no right to a flow of air as such; so an adjoining owner could disturb the flow or even divert it, eg by a building or a plantation of trees.

However, if the building had a chimney which polluted the flow of air, the disturbed landowner could take action.

Drainage

The freeholder has the right of natural drainage. Thus, a neighbour on lower ground must allow the free movement of natural water from the freeholder's land. This does not apply to water draining directly from buildings unless an easement has become established.

Support

Land in its natural state, ie without buildings, has the natural right of support. Thus, the owner of adjoining land must not do anything which would cause the land to collapse, eg into a hole dug on the land.

Water

A landowner's water may be a natural pond, lake, spring, stream or river. His natural right in water on the land has four features:

1. The accustomed flow of water in a natural stream or river on the land
2. The water be unpolluted
3. The fishing in the water on his land
4. The use of a reasonable quantity of water from a natural stream or river for purposes associated with the land, ie riparian purposes such as watering, but the quantity must not diminish the accustomed flow of other downstream landowners.

Customary rights and the like

Rights to a green

Similar to rights to a common, rights to enjoy a green or village green are ancient rights which must now be registered.

Public footpaths, bridleways and the like

Over the centuries numerous public rights of way (footpaths, bridleways, etc) have been created by:

- ancient customary use

- a landowner's dedication or gift of land for public use; and
- action by a highways authority in creating such a highway.

Profits or profits à prendre

Profits have their origins in the rural feudal economy of the Norman Conquest (and, no doubt, even before) — typically they are rights in common and ancient forest rights. They allow a particular person or group of persons to enter on to the land of another and take something of value, eg timber.

While it is not necessary for the right-holders to have land, they may personally own a so-called dominant tenement. In ancient times these rights were recorded from time to time, so the particular groups of persons who enjoy them are known, in the sense of being within a class or group, eg the owners of, say, ancient cottages near certain woodlands. (They are not rights available to every member of the public.)

Profits may be several, ie enjoyed by an individual or common owned by two or more. Examples of these rights include:

- taking of coal, gravel, sand or the like, ie part of the land itself (soil)
- cutting of peat or turf as fuel (turbary)
- taking of fish (piscary)
- grazing of cattle (pasture)
- cutting of bracken
- taking of timber (hay-, house- or plough-bote)
- grazing of pigs (pannery).

Some rights may be local, eg there are ancient rights to minerals in some areas — lead in the Peak District and iron in Dorset.

Creation of rights

Rights are created in the following ways:

- grant
- reservation
- squatting and adverse possession
- prescription; and
- statute, eg statutory wayleaves and easements and certain prescriptive rights.

Grant of a right

A freeholder may grant a right to another. This often happens when land is sold and the seller, for instance, grants an easement of way across the retained land for the benefit of the land taken by the buyer.

Reservation of a right

Somewhat similar to a grant, a freeholder who sells land may reserve a right over the land being sold to another for the benefit of the retained land, eg a right of way.

Prescription

A landowner may obtain a right over the land of another by unauthorised use of the land in some way for a specified period. The creation of such rights had their origins in:

*   common law
*   a statute, the Prescription Act 1832 which governs their creation.

For the common law right, there are several approaches; all of which are subject to conditions.

*   First: the user must show continuous use for at least 20 years and, therefore, its presumption of use since 1189. The owner of the alleged burdened freehold land may be able to show that was not possible and hence rebut the common law claim. A successful claim would establish a prescription at common law.
*   Second: lost modern grant is another example. Here the user, say, the owner of adjoining dominant property, has used a way across the alleged burdened land for many years, as had predecessors to the dominant title. The owners of the burdened land seek to stop the use but the court may well presume a grant of the way to made to a previous user (then owner of the dominant property) — this is the presumed lost modern grant, so-called.

For the 1832 Act there are three situations:

*   prescription by 20 years use: a claim after 20 years of use (other than a use of light) will succeed if the defences can be overcome
*   prescription by 40 years: a similar situation exists for 40 years of use but there are fewer defences that the owner of the burdened land may adopt
*   prescription of a right of light: under the 1832 Act a right of light may be claimed after 20 years and the claimant is not limited to freeholder owners.

There are defences against a person obtaining a right of light, eg by placing a hoarding on the servient tenement outside of the aperture on the dominant land. However, the burdened owner may, within the 20-year period, use the Rights of Light Act 1959.

Right of Light Act 1959

The Act gives an effective a notional practical blocking of the light which has yet to be claimed. Prior to the 1959 Act the burdened owners were required to erect a hoarding in front of the aperture through which the light was enjoyed. Since the Act a notice of a notional hoarding (so to speak) may be registered to the same effect.

Ancient lights

So-called ancient lights are very old apertures or windows, eg to a listed building, where the right of light almost certainly exists and a defence against a claim is almost invariably unsuccessful.

Trespassers, squatters and adverse possession

A peril which the owner of a vacant property faces is the possibility that it will become occupied by squatters. Until at least 12 years of possession are completed, they are trespassers. As trespassers the owner may seek an eviction order for possession against them from the county court.

Formerly, the 12-year rule of adverse possession under the Limitation Act 1980 enabled squatters in possession for at least that period to obtain legal ownership against the owner, even those with registered title. The Land Registration Act 2002 changed the law on registered land in that a squatter cannot succeed against the owner of land with registered title.

Statutory wayleaves and easements

Sometimes a company or public body needs to lay a pipe or cable underground on someone's property. Other bodies may want to erect poles and string wires across the property. The rights embodied in these actions are called wayleaves and easements — they have a place in society under statutes.

Thus, many statutes give utility companies, statutory undertakers and others the right to compulsorily place a pipe or cable on, under or over land. The statutes have each generated their own mini-body of practice and concomitant case law, covering such facilities as:

- electricity
- gas
- telecommunications
- sewers
- water
- oil and other pipelines.

The owner who suffers inconvenience, loss or damage may claim compensation or annual payments. Over the years a substantial body of case law has developed on the measure of compensation for the various types of pipe, eg gas, sewer and water, or cable, such as electricity cable or national grid oversail.

Sometimes, annual charges are agreed jointly by the utilities' representatives and the landowners' representatives to reflect the nature of, for example, electricity poles of various shapes, eg A-poles and H-poles. Sometimes a negotiated capital sum is paid, eg 50% of the capital value of the acreage, when a gas trunk line or oil pipeline is laid. This will be paid with any other items of claim which are agreed on a case by case basis.

Sporting rights

Sporting rights, ie the right to fish, shoot and hunt, attach to the ownership. The owner has several options, namely:

- enjoy them as of right with family and friends on a non-commercial basis (subject to any restrictions under statute)
- as above, but also issue short term licences on a relatively casual basis
- create and operate a business entity to commercially exploit the sporting rights
- sell some of the land with the sporting rights attached
- sell the sporting rights outright (but the land is retained)
- grant the sporting rights to another individual or body.

Taxation

Value added tax (VAT) and income tax may need to be accounted for in some of the revenue generating options. Similarly a disposal of land with the sporting rights attached may result in capital gains tax and, possibly, VAT on an apportionment of the consideration (sporting rights are standard rated).

Rights of access or entry

Amenity, planning, environmental and other officials

Enactments empower officials to enter land and buildings in the course of their work. If access is improperly denied by the occupier a penalty may result. Nevertheless, except in emergency, adequate notice is normally required and the official should proffer identification. Box 4.2 lists many of the officials who have rights of entry for a particular purpose.

Box 4.2 *Officials and others with a right of entry to land and buildings*

Officers of an acquiring authority	• to inspect for the purpose of valuation
Valuation officer	• to value for rating or for council tax
Officers of Inland Revenue and Customs and Excise	• to examine the papers of a business for the purposes of tax enforcement
Planning officers	• to seek evidence of non-compliance of the planning regime
Building control	• to inspect on-going construction work for compliance with the building regulations
Police officers	• in emergency situations or in tackling crime (may need a warrant)
Local authority officer or other authorised person	• entry on neighbour's land to deal with a high hedge under s 74 of the Anti-social Behaviour Act 2003
Landowner	• to effect repair and maintenance in accord with the Access to Neighbouring Land Act 1992
	• entry under the s 8 of the Party Wall etc Act 1996 to effect repairs
Treasure seeker	• person with the permission of the owner, enters seeking objects with a metal detector
	• on making a find, the individual must comply with the Treasure Act 1996
Ramblers and roamers	• ramblers keep to public footpaths (but may enter other land with the permission of the owner)
	• roamers enter land which is open countryside

High hedges

The Anti-social Behaviour Act 2003 provides right of entry to local authority representatives to deal with high hedges.

Neighbouring land

A landowner may obtain a court order for right of access to carryout essential repairs and maintenance under the Access to Neighbouring Land Act 1992.

Right to roam

The right to roam in open country was provided by the Countryside and Rights of Way Act 2000. Although much land is covered by the right to roam, there are about a dozen groups of excepted land which must not be encroached, including:

- aerodromes
- certain lands regulated under the Military Lands Act 1892 and 1990
- golf courses
- mineral workings and quarries by surface working
- land of statutory undertakings
- land covered by the telecommunications code.

As yet the right to roam does not extend to coastal land but there is power in the Act to extend the right to roam in this respect.

Extinguishing or diverting rights

Generally, rights are extinguished when one of the following happens:

- abandonment by the owner or owners of the right
- merging of property ownerships, eg the owner of the burdened land buys the dominant land
- release, implied or express, by the owner of the right
- process of law, eg compulsory purchase.

Some of these are straight forward but others require an explanation.

Compulsory purchase

On the compulsory purchase of land which is subject to a right belonging to another, the acquiring authority may:

- allow the right to continue, or
- effectively acquire the right and so merge it with the acquired land.

In the latter instance, the owner of the right may be entitled to compensation under s 10 of the Compulsory Purchase Act 1965. Typical situations include:

- an easement
- a restrictive covenant.

Footpaths and other highways

Highways legislation gives highway authorities the power to close or divert a footpath, bridleway or other public highway.

Restrictive covenants

A freeholder wishing to sell or lease a site for development (or the prospective developer) may find that the land is subject to a restrictive covenant which either prevents development or hinders the full scope of the site. Some might be tempted to proceed; prudently some means of allaying the fear that an owner of right will appear is needed. Possibilities are dealt with in Chapter 17.

Wayleaves and easements

Utility companies and statutory undertakers holding statutory wayleaves and easements which burden land will normally co-operate with the landowner who requests a diversion or removal.

Safeguarding against rights

Statutory declarations

An owner, who is burdened by the use of a way on land which is not yet dedicated to the public, may retain the private status of the land by making a statutory declaration every six years; in effect giving permission to the use of the way but retaining ownership rights.

If the land is used without the authority of the owner for at least 20 years the highway authority may declare it as dedicated to the public as a highway.

Insurances

Insurances are commonly used for seemingly abandoned rights, including restrictive covenants. The landowner who wants to carry out development which may contravene an old, but seemingly abandoned, right would be prudent to obtain insurance against any loss arising should the owner of the right successfully challenge the contravention.

Demolition and clearance of land

When land is cleared or buildings are demolished, care should be taken that the rights of others are not contravened by the developer. Some of the issues are dealt with in Chapter 18.

Part 2

Property Markets

Property Marketing

Aim

To understand marketing in the property industry

Objectives

- **to examine the term "marketing" in the property industry**
- **to describe each marketing function in terms of property**
- **to illustrate marketing to a particular group**

Introduction

Three descriptions of the term marketing seem to be used by those in the property industry.

1. To describe the role or cluster of activities of estate agency.
2. Within estate agency, to describe a somewhat narrower cluster of activities, eg advertising, undertaken when a property is promoted to those in the market.
3. To describe a procession of functions in marketing of property, including what are commonly known in business management as:
   - market research
   - market segmentation
   - product development
   - channel development
   - pricing
   - financing
   - selling
   - promotion
   - customer care.

None of the three descriptions is wrong in its context but the last will be used conceptually to explore the nature and scope of the property market. Estate agency and hence advertising and other promotional activities are dealt with in Chapter 12.

Roles and activities

Box 5.1 gives the main roles and activities in the property market.

Property and society

As indicated in Chapter 1, land is a unique resource. It has to provide for all basic needs and, perhaps more besides. Developers of property and other suppliers of property-related goods and services are interacting with a population that is continuously changing its pattern of demands for accommodation of all types. In principle the changing pattern is being discerned by market research (and by intuition, perhaps) in such ways that developers are encouraged to specify and procure buildings or sale or to let to meet the demands of specific groups (market segments).

Market research

The term market research is taken here to mean research which investigates some aspect of a property sector, such as agricultural business tenancies or self storage facilities. Marketing research has a different perspective in this volume. It is taken to be a broader kind of study area, eg the conduct of marketing functions or relations in more than one industry. It is, therefore, likely to be of less interest to those in the property industry.

As a result of the research for a proposed development, a developer should be in a position to understand the segment of the market and to formulate a marketing mix for the expected buyers or tenants.

Market segmentation

Market segmentation is a process by which the market is investigated with the view to identifying the needs of those in the market for a type of property. Upon the completion of a study, it should be possible to decide whether there is a case for proceeding with a property development since the results should give the number in the market who can afford the marketing mix which is to be offered — such that the developer may expect to profit.

Marketing mix

When a unit of property is offered on the market the developer may have a clear notion of the applicant who is expected to be interested. The developer will provide or offer a specific cluster of features to go into the offering of the physical unit of property. It is intended to attract the right applicants, and constitutes a marketing mix. Every marketing mix for property is likely to contain all or most of the following:

Box 5.1 Roles and activities in property market

Owners	• owns property • buys and sells property • develops property
Occupiers	• either an owner or a tenant • either lives in, runs a business in the property or both
Estate agent	• advises on the sale or letting of a property • acts for seller to promote property for sale or to let
Auctioneer	• advise on the suitability of the property for sale by auction • acts for seller to sell property by auction
Valuer	• estimates price of property that is for sale or to let • evaluates and advises on condition and marketability of a property
Building surveyor	• carries out surveys • designs relatively small buildings and works
Home inspector (see Housing Act 2004)	• will carry out inspections for the forthcoming housing condition report (HCR)
Solicitor or conveyancer	• prepares contract for sale and completion
Land surveyor	• surveys land and prepares plans and maps
Land registrar	• registers land ownership • guarantees the entry in the register • empowered to register detailed boundary location plans
Mortgagee	• lends money on the security of the property
Insurer	• covers the property against fire, flood and other perils
Local authority	• maintains local searches register • bills or charges council tax and business rates and collects, by enforcement if necessary
Ombudsman for Estate Agents (OEA)	• hears complaints about estate agents conduct and adjudicates • may order up to £25,000 in compensation • has published a Code of Practice • logo indicating a good trader has been instituted for member estate agents • probably collect data for the home information pack (HIP) expected 2007 (see Housing Act 2004)
Property researchers	• research aspects of the market on a commissioned basis or within a consultancy
Runner	• individual who seeks information about possible or likely property transactions in prospect • acts as a gatekeeper of the information in the market
Journalists/publishers	• takes advertising or free publicity features • publishes books and articles

- promotional mix, eg advertising, brochures and free publicity
- viewing service through an agent
- tenure, eg freehold, leasehold, timeshare
- suitable accommodation, eg space size, quality, rooms and general services
- specific services for particular occupants and visitors, eg serviced offices and self storage space
- price, eg in terms of capital sum, rent and rent or premium
- service charge for landlord's services
- finance, eg for purchase, adaptations or fitting out.

The heart of marketing and this chapter is the marketing mix for property.

Government intervention

At times the government intervenes in the property market. There are numerous examples which are cited throughout this volume. For instance, Chapter 14 examines the notion of acceptability and sustainability of settlements and buildings. These are mainly concerns about quality control of buildings which the government imposes. Of course, developers now adopt the notions of sustainability as part of their promotional mix.

Product development

The product is the building together with terms and conditions on which a particular property or building is to be sold or let. On the one hand, product development may be construed as the processes from designing the property for its site, its construction and handover (see Chapters 13 to 19). On the other hand, it may be considered to be only the design part of the process — a conceptual realisation. Construction may result in design changes but these are not likely to be significant.

Channel development

Channel development is a continuous process of trying to improve the way in which a market operates. In the marketing of property it is a gradual process with several groupings of organisations. For the non-professional buyer or seller, the channel is largely hidden unless an individual has previous experience of buying or selling. Even then many of its features will be unknown. Similarly, for some the notion that a continuous process of improvement or attempted improvement, ie channel development, is happening may seem far fetched.

Government

The government has improved or been pressing for improvements to the channel of residential estate agency by:

- establishing on a voluntary basis the ombudsman scheme for estate agency
- reforming land registration procedures with the introduction of electronic conveyancing (under the Land Registration Act 2002)

- trying to get acceptance of the home information pack, now embodied in the Housing Act 2004.

Professional bodies

The several professional bodies concerned with estate agency have worked jointly or severally on channel development for estate agency for very many years. Elements include:

- special groupings so that their members may address , among others, channel development issues and concerns
- advice to government and others seeking to improve estate agency
- developmental training for estate agents
- clients' accounts
- professional indemnity insurance
- complaints procedures for alleged misconduct
- e-estate agency
- codes of conduct
- continuous professional development schemes and training sessions.

Estate agencies and individual estate agents

Over the years estate agencies and individual estate agents have developed new practices and technologies to support the role, including:

- reassuring clients and applicants, eg with a complaints procedure and membership of the ombudsman scheme
- endorsing their practice with corporate or individual membership of recognised and respected professional bodies
- building or joining internet property portals for the promotion of clients' properties
- internet portals for the promotion of property and property services
- creating and joining a training organisation for estate agency
- forming and joining a professional body for residential lettings which is active in channel development in that field.

Ombudsman for Estate Agents

There are about 12,000 offices of residential estate agents with about 6,000 who are members of the Ombudsman for Estate Agents (OEA) scheme. At present membership is voluntary but the OEA wants the principal professional bodies to make the scheme compulsory for their members.

Pricing and financing

In the property industry pricing refers to the processes of setting prices for sale or leasing property.

Capital and rental payments

Pricing ranges from capital sums for freehold sales to annual rent for leases and tenancies; in some instances the latter will have a lower rent together with a capital sum. In the open market, supply and demand are the determinants of price but pricing is not always a matter of open market forces; the government has frequently arranged the basis of pricing rents, eg the former controlled rents and regulated rents for tenanted housing. On capital values various statutory codes are used to deliver government policy, eg the price payable for houses on long leases (under the Leasehold Reform Act 1967) and the open market value for compulsory purchase (under the rules and assumptions of the Land Compensation Act 1961).

Inclusive payments

When comparing different properties care should be taken to clarify the precise terms for a lease. Thus, occupying tenants may regard as evidence of high rents those properties which are held on a lease at:

- a rent inclusive of repairs, maintenance and insurance or
- a rent inclusive of charges for services or
- a rent inclusive of rates.

Here, however, the landlord's offering in each instance is a different marketing mix. The landlord is bearing costs which the tenant would otherwise pay.

Financing

The developer will normally want to finance the purchase of land and the construction of a development. For most buyers or lessees, the developer's involvement in financing their transaction seems usual. Nevertheless, some developers may offer special terms with a bank or financial institution, eg a mortgage. Also, the offer to lease a property is a form of zero-financing as far as the tenant is concerned. Similarly, a developer's offer to exchange the buyer's existing property for a new one is a form of financing. These examples are part of the marketing mix.

Selling and promotion

Selling is the process of interaction between the seller (or the seller's representative) and the prospective buyer. It comprises:

- attracting the applicant
- informing the applicant of the property's features
- showing the property to the applicant
- negotiating the price with the applicant
- closing the agreement
- as necessary, progressing the sale or letting to contract and completion.

Promotion — purpose

When property is offered for sale or to let, promotional activities are undertaken, ie prospective applicants are informed, encouraged to view and to make an offer by promotion. Promotion is intended to inform a potential those in the market, eg an applicant as follows:

- of the many attributes of the property's marketing mix — a basic information
- that the bundle of attributes suits the applicant's requirements — an emotional appeal perhaps
- of the key features of the buying process (as conducted by the seller or a representative of the seller) — creating an understanding of the process
- of the positiveness of any endorsements of the property's marketing mix — again an emotional appeal based on any third party support.

Promotional mix

A typical promotional mix may include:

- advertising — paid advertisements placed in newspapers, magazines, and journals
- publicity — free editorial and features about the property placed in newspapers, etc or other material which is distributed in the market
- managed events — information distributed at exhibitions, conferences and other events by brochures, leaflets and posters
- personal selling — details given by word of mouth to key market gatekeepers of information.

Target audiences

It is not sufficient to prepare the promotional mix and hope that it has the desired effect. A target audience will need to have been identified in terms of the following:

- buyers — those who provide the wherewithal to pay for the property
- users — the one or more persons who will occupy and use the property, either as residents, as management and staff of a business or as some other user, eg a customer in a club
- influencers — those who have a say in the appraisal of the property and the decision to buy
- deciders — the person or persons who decide whether to buy.

It follows that their message needs are not quite the same in terms of information, slant, emotion, etc.

Customer care

Customer care is the term which describes the activities the developer, contactor or landlord undertakes to maintain good relations with the buyer or lessee who has taken property.

Leased property is a joint arrangement in which customer care by the landlord may be best evident. Property management or facilities management are the functions in the property industry where customer care is likely to be evident or not. In construction, snagging is an example of customer care, as is the latent defect insurance which is offered with new buildings.

Property markets

The property market is complex and, in effect, comprises the markets or sectors of many different types of property, eg the industrial property. The participants in each property sector are buyers of space either as owners or as tenants (paying rents or premiums). The market is, therefore, a multi-sectored exchange. The owners are those who hold property in a general sector, namely public, private and voluntary (each of which may be sub-divided). The general sectors are:

- private sector buyers and sellers, each with one or more particular objectives (dealing, investing, business, residential and leisure — see Chapter 6)
- public sector buyers and sellers with changing statutory duties and discretionary powers (many have buying power based on compulsory purchase)
- voluntary sector buyers and sellers, many with social or charitable objectives.

The property sectors are usually categorised as commercial, industrial, residential, leisure and so on. Most sectors have niche markets, eg self-storage within the distribution and storage sector. Business or domestic customers of a typical self-storage centre are offered a developed channel, format or model with the following marketing mix set out in the case study.

Marketing mix — case study

Box 5.2 Marketing mix for a self-storage unit	
Target customers	• business users needing space in good location with up to 24 hours access • domestic customers with inadequate space or needing space when on the move to a new home • students with short term space needs
Space	• storage space may be hired for rent on a monthly or daily basis • size of the space is variable, starting from say a m³ to 100 m² or more • space is at room height or more
Payment	• payment can be by cash, cheque, credit card and direct debit
Concessions	• students or others may be offered concessions on rent and insurance
Security	• the accommodation is secure with CCTV monitoring, out-of-hours monitoring and other systems • accommodation is locked with hirer's own padlock • padlock may be purchased
Access	• access is 24 hours a day
Handling goods	• trolleys and other handling equipment is available • storage boxes of different sizes and types, eg for dresses, are available
Insurance	• cover starts at, say, £2,500 minimum
Removers	• external removers, if needed, can be recommended

Promotional messages	Business customers who do not otherwise own or lease property: • goods delivered to premises by suppliers and safely await storage by business customer • good location for business travel • need for flexible space requirements can be satisfied Domestic customers: • temporary space needs whilst moving house can be satisfied for short period Students: • leave personal belongings while away on long vacation or on field trips Motorists: • location is near the road the motorist is on Television watchers: • there is a unit in your area
Location	• tend to be in urban areas near motorway intersection or on or visible from a major road

Property Owners and Others

6

Aim

To establish why property is owned and who owns it

Objectives

- **to describe the seven prime objectives for owning land**
- **to show the importance of legal personality in owning property**

Introduction

The owners of property usually have one of several main objectives when buying or renting a property or a portfolio of properties. This chapter explore the concept of the objective of ownership (of property). First it briefly examines the purpose of ownership and activities relating to assets in general with emphasis on the financial aspects.

Asset ownership and risk

Most individual hold tangible and intangible assets from which they seek to obtain two benefits, namely:

- an actual or notional annual net income
- increase in the capital value of their investment over the period of ownership.

They may measure their success by calculating the annualised total return and comparing it with the cost of borrowing capital or some other relevant benchmark.

In a sense, provided the total return is obtained, it could be said that the individual should not be concerned about the nature of the asset, ie it could be property, shares or some other thing. In practice, the individual's concerns do not hinge around the one measure (total return) but are about a number of factors which will continuously underpin actions attempting to achieve the desired outcome. For instance, in a bear market, an individual investor with a portfolio of shares will take actions to minimise the losses, perhaps by switching to an alternative field of investment, eg property. In the same position another investor may decide to ride out the bear market in shares — taking a longer perspective on the holding. A third may choose to liquidate some shares and put the proceeds into fine wines and a collection of stamps. The issue is the about the principles which underlie the choices.

Essentially, the individual will probably feel safer (in that risks are understood, measured and balanced in the decision-making) when giving attention to the following:

- hold assets in a number of fields — the diversification principle
- hold assets which are tax-favoured — the tax avoidance principle
- hold assets which require the minimum of attention — the reduced management or self-managed (but monitored) principle
- hold assets where loss or damage may be borne by others — the risk transfer principle, eg fire insurance
- hold assets which enhance one's self to self, and to others — the status-moral-status or spiritual principle.

Of course, the above does not suggest that all are alike — some are high risk-takers and so on — merely, that whether knowingly or not, all decisions could be said to follow the above.

Objectives of owning property

A person, ie a legal persona, usually has at least one objective or *raison d'être* for an acquisition of property. Broadly objectives may be divided into private, public and voluntary sector ownership; although transfers between the sectors happen, eg on privatisation when a publicly owned water utility becomes privately owned.

For a person in the private sector, primary objectives may be described as follows:

- dealing
- investment
- residential occupation
- business occupation
- leisure occupation.

An owner may have a mix of objectives in acquiring a property. For instance, a young person may buy a flat for owner-occupation but sub-let part of it (investment objective) to assist with servicing the mortgage which was obtained to buy it. Similarly, the board of a company set up as a property dealing company may decide to retain some assets for investment purposes. It may need to restructure itself as a group of two or more companies, at least one of which is a property investment company.

In the public sector the objectives fall into three conceptual groups, namely:

- functional — where the main purpose is to provide a service, eg a sewerage pumping station, indirectly to local residents or businesses
- subjective — where the organisation occupies the building
- objective — where the main purpose is to provide services directly to those who enter the property, eg a school for school children or a hospital for the sick.

Box 6.1 lists the types of owner by objective and briefly indicates their characteristics.

Box 6.1 Objectives of ownership of property

Dealing		
	• seeks profit from sale	• income taxation on profit
	• short-term ownership	
	• short-term funding	
	• re-iterative process	
	• often does work to property before selling it	
	• income taxation on profit	

Investment
- seeks rental income
- seeks capital appreciation

  • income taxation on net rent
  • capital gains tax on gain

Residential
- location for employment
- location for education
- property is fit for life-style
- desires capital appreciation
- costs of running not excessive
- issues of local taxation

  • stamp duty land tax likely on purchase
  • capital gains tax exemption except in certain instances
  • council tax liability in Great Britain
  • rates liability in Northern Ireland

Business
- image fits the corporate style
- location for market
- location for raw materials and supplies
- location for staff
- transport is appropriate
- communications are appropriate

  • liable to income taxation on notional rental value
  • liable to capital gains tax
  • stamp duty land tax on purchase
  • liability to business rates
  • climate change levy may apply
  • value added tax may apply

Leisure (individual or family)
- fit for life-style enjoyment
- communicates life-style
- non-business objective

  • stamp duty land tax on purchase
  • liable for capital gains tax on disposal

Public
- functional

- objective

  • sewage works
  • laundries

  • schools
  • leisure centres
  • day centres
  • hospitals and clinics
  • zoos

Voluntary
- objective

  • orphanages
  • hospices
  • nursing homes

While an owner will have one or more of the objectives given, it is important to appreciate that the owner's legal personality has a bearing on ownership and other property matters. In the next section a series of profiles set out the main characteristics of different legal persons.

Record of ownership profiles

Although it may seem surprising, occasionally an owner has attempted to sell a property which it did not own. Similarly, an architect was asked to design a conversion of a leased part of a larger property. Subsequently, it was found that the leased accommodation was actually smaller than the part specified for the design. Finally, a large business organisation was intent on selling its owner occupied multi-property estate on a sale and leaseback basis. However, before it could do so it had to check and confirm the ownership of each property in the estate. The problem was that over many years the records had not be properly maintained; it is possible that they were not designed for a large-scale disposal. The lesson is that appropriate estate records need to be established and maintained by the estate owner, thus ensuring quick access to every property's:

- legal profile
- accommodation and services profile
- occupation profile
- works profile
- budget (annual and capital)
- business plan.

Property and the owner's legal personality

Legal personality or status describes or contains the bundle of rights and obligations attaching to an individual (as such), company (as such), a trust (as such) or other body. Every partnership, company and other body is set up under particular legal documents which describe and prescribe what the body is permitted to do with its property. An individual is a legal person (see Box 6.2). Other legal personalities that an individual may want to create in association with property include the following:

- a sole trader
- a partnership
- a limited company
- a group of companies
- a trust (see Chapter 2).

Their main characteristics are shown in Boxes 6.2 and 6.3. There are other private sector legal persons but these are companies or other bodies which are essentially similar but larger. Other property owners in the public and voluntary sectors are explored below and in Chapter 32.

Criteria for selecting a property

Property must suit the objective (or possibly a mix of objectives) and status of the prospective owner. Thus, each objective seems to induce different perspectives or criteria in choosing property, eg time scale.

Box 6.2 Some of the characteristics of an individual to conduct business and hold property

Individual (adult)	• may buy and sell as a resident • may enter into contracts to insure, improve, etc property • may buy and let property • may become a sole trader and form partnerships and companies	• taxed to income tax on all income sources • estate liable for inheritance tax on death
Individual (unsound mind)	• rights constrained — property held on trust on individual's behalf	• as adult
Individual (minor)	• rights constrained — property held on trust on minor's behalf • contracts are voidable by minor may be the beneficial owner of property through trustees	• taxed as an individual adult (some relief)

Box 6.3 Some of the business legal entities and their nature

Sole trader	• may run a business as a sole trader, possibly using personal name • may run a business with a registered business name	• sole trader sets up with minimum of formality • business essentially private • no publicly available accounts • regulation is less onerous • taxed to income tax and inheritance tax (see Individual)
Partnership	• partnership conducts business as legal entity • profits shared according to the partnership agreement	• partners jointly and severally liable for the debts of the partnership • governed by the Partnership Act 1890
Partnership (limited)	• partners taxed as individuals on their share of the profit	• partners have limited liability • limited partnerships allowed under the Limited Liability Partnerships Act 2002
Company (private)	• company conducts business as legal entity • company is not listed on a stock exchange	• shareholders' liability is limited • tax on dividends is borne by the shareholders • company is liable for corporation tax
Company (public limited company)	• shareholders' liability is limited • company is listed on the main stock exchange, the Alternative Investment Market or the Ofex market • company is liable for corporation tax • shareholders bear income tax on dividends	
Company (limited by guarantee)	• company is a corporate vehicle commonly used by housing associations	
Community interest company	• a company set up under the Companies (Audit, Investigations and Community Enterprise) Act 2004 • activities are of benefit to the community • must pass the community interest test	

Time scale

When buying property different people have different ideas about how long they want to own or occupy it. Much depends upon what they intend to do with it. The person who sees the opportunity to buy a property as an opportunity to profit from a re-sale will invariably have a different perspective from one who wants to occupy or let it for a longer period.

Box 6.4 merely suggests the property owner's likely view of the period during which the property will be owned.

Box 6.4 Duration of ownership		
Dealer	short term	• when sale would achieve a profit • subject to market conditions • subject to marketability • until repair, maintenance or new build works are complete • until a buyer is found
Investor	long term	• until investment no longer meets investment expectations • until standards on rent and prospective growth are failing • until a better investment opportunity arises • where the property is part of long term redevelopment proposal
Resident	short term (say 3 to 5 years)	• parent who buys dwelling for son or daughter at university • buys on basis of a temporary stay in area • first time buyer or footloose young professionals wanting to move up market or away • elderly downsizing or moving into care
	medium term (say 5 to 10 years)	• young family wanting to move to larger home • downsizer, until elderly and then moves into care
	long term (say 10 years to end of life)	• established family or elderly lone spouse (until downsizing) • stately or family home owner (intending never to move) suffers hard times
Business	short term	• occupant of serviced offices • occupant of business self storage unit • start-up business on a short lease • lessee of premises in redevelopment site
	medium term	• business on a medium term lease (may have right to renew the lease) • growing business on a constrained site • failed or ailing business where the shareholders, creditors or mortgagees may force a sale
	long term	• long established successful business in a particular location • where the property is part of the business's goodwill or is prestigious, eg an HQ building
Leisure	medium term	• mobile home ages and must be removed from leisure site • need fades as family life cycle progresses, eg decline of usage of young-child-oriented leisure facility • management of, say, a timeshare or a camp site allows the facility to decline below acceptable standards

Liquidity

The ease with which a sale is achieved is referred to as liquidity. A hard-to-sell property may, of course, take years to sell. Normally, an owner who puts a property on the market would expect a sale within a couple of months. Dealers would want buy and resell property rapidly.

A comparison with quoted shares is fruitful here since shares can be sold through the shareholder's stockbroker in an instant either electronically or by telephone. In fact shares in a quoted property investment company with a property portfolio may be seen as a means of avoiding the liquidity problem associated with directly owning a similar property portfolio.

Various proposals have been made to change ownership rights in individual properties by creating ownership rights in shares in the property. Ideally, the person who owns all the shares in the property owns the property. A subsequent disposal of all or a portion of the shares to others eases the liquidity problem of the property owner. A market for individual property shares would probably need property stockbrokers and a range of services similar to those available to buyers and sellers of company shares. However, this kind of framework remains conceptual at present.

Wealth retention

Traditionally, property has been a basis for holding and accumulating wealth. The selection of an individual property obviously depends upon the initial capital which is available or can be raised, nevertheless various means are used to achieve the bricks and mortar effect, including:

- letting property to tenants with good covenant
- improving the property by development, extensions or makeovers
- buying adjoining property to garner a marriage value effect
- buying other interests in the same property so enjoy the marriage value or special purchaser effect
- buying or inheriting property which enjoys exemption or relief from taxation, eg favoured assets (see Chapter 27).

Property protection and risk management

Any property is subject to risks which should be taken into account on selection. They include:

- action by public bodies, eg compulsory purchase
- the prospect of taxation liabilities (see Part 8)
- use of insurances (see Chapter 22)
- security measures commensurate with perceived risks
- property rights of others, eg restrictive covenants (see Part 1)
- careful selection of professional advisers and contractors (see Part 3).

Dealers

Some individuals or companies become owners of land or properties in an effort to profit from a number of somewhat similar activities, including:

- acquisition followed by a resale of the property without any improvement
- following acquisition of a site, the construction of roads and the division of the site into plots, ie lotting of the new frontage land; each new owner of a plot carries out its development
- as previous, but the developer carries out development and sells each property on completion
- acquisition, development or makeover of an individual property followed by the sale of the improved property.

On the face of it they are seeking to make a profit from whatever kind of activity they undertake. Long established case law has given criteria, the so-called badges of trade, for determining whether the kinds of activities described above are examples of dealing. Thus, those who are in business as house builders are, in effect, dealing in property. Generally, it is not now a problem to recognise dealing but new situations do arise from time to time to keep the courts busy.

Taxation

All dealers are assessed under schedule D case I of the Income and Corporation Taxes Act 1988. It has been or is sometimes difficult to decide, particularly for taxation, when a person is dealing. For instance, an individual purchased a house and improved it prior to sale and then repeated the activity several times (living in each property in turn). Dealing may have taken place but not necessarily — if the activities took place within a one-year period it is almost certainly dealing. However, if the period was 35 years, it is unlikely to be dealing. For taxation, the badges of trade, derived from case law, are indicators of the nature of the objective. The badges include:

- re-iteration — repeating an activity several times
- documentation — the correspondence may reveal intention to trade (or invest, etc)
- formal status — legal status may indicate dealing, eg a company's memorandum of association, etc
- works to property — carrying out particular works may suggest intention, eg converting a building into flats
- funding — short-term funding or funding at a deficit may suggest an intention to trade.

Dealers do not enjoy capital allowances directly, nor do they pay capital gains tax on gains on their property which is held as stock-in-trade.

Funding

In principle, dealers do not necessarily require long term funding in that the first profit pays for the next with short term borrowing; thus, a cash-generative business could well enable the repayment of any debt and the funding of future operations, including capital expenditure. In the property industry, a housebuilder's treasury function is like to seek long term equity funds or debt to build up a land bank and to maintain on-going operations.

Investors

Some investors buy the completed property while others buy land, build property and let it, retaining the property in their portfolio. The wide ranging property investment sector comprises:

- insurance companies
- landed family estates
- banks
- insurance companies
- the Crown Estate and the Church Commissioners
- property investment trusts of sectoral interests
- a multitude of individuals, perhaps as directors or trustees of family various legal entities.

Funding

Investors tend to hold property for a long time, so funding on a long-term basis is the norm, eg by mortgage. Also, they are unlikely to use deficit funding. However, developers may construct a development with short or medium term funding and the interest is rolled up on the promise of a mortgage. Deep discount bonds with a lot coupon may be used for a given period — on the expectation that the rents on future lettings will cover the interest (and any deficit of interest). In due course the rents may be expected to cover the commercial going rate of interest.

Taxation

Investors in property are assessed to income taxation under schedule A of the 1988 Act. Generally, they are able to enjoy capital allowances on eligible expenditure on qualifying buildings. Liability to capital gains tax arises on chargeable gains (see the Taxation of Chargeable Gains Act 1992).

Residential owners

Owners of residential property are looking for life-style attributes in the property. The property will often be linked to the stage of life of the owner or owners. Generally, buyers may be divided into market segments:

- the first time buyer on a relatively low budget
- the young professional early in a career but with a regular earnings of good basic salary and, perhaps, commission
- the young family
- the maturing family with teenagers
- the older person or couple in a house which may be too large
- the older individual or couple who down-sized to a bungalow, flat or sheltered accommodation
- the self-build developer wanting a home
- the older individual in a care home or nursing home.

Taxation

Apart from council tax, residential owner occupiers do not, generally, pay income tax on the notional rental value of their homes. There is the view that if home owners give the property to, say, one of their children and remain in occupation without paying rent, the Inland Revenue will be able to charge income tax on the annual gift rent they enjoy, notionally payable by themselves! Also, most owner occupiers do not pay capital gains tax when they sell their sole or main residence. However, if the property has been part tenanted or has a large garden, a liability may arise.

Owner occupiers escape income tax on rents up to a certain amount under the rent-a-room provisions. Box 6.5 shows the tax situation of the owner of a house who puts it to one of several various uses.

Box 6.5 Taxation on the uses to which a house may be put

Owner occupation
- family residential use only
- family use with room for owners' business

- with lodger on rent-a-room

Let to tenants (unfurnished)
Let to tenants (furnished)

Boarding house

Small hotel

Holiday cottage
Second home for family

- no income tax
- income tax on business
  - may claim a portion of the expenses of house
- no income tax as long as rent within limit for the year
- income tax on net income
- income tax on net income (allowance for furniture)
- income tax as a business

- income tax as a business

- income tax as a business
- no income tax

- no CGT unless a large garden
- CGT liability may arise (particularly where expenses claimed
- no CGT on that account

- CGT liability
- CGT liability

- CGT (part family occupied may be treated as exempt
- CGT (part family occupied may be treated as exempt)

- CGT liability
- CGT unless election as main residence

Business owners

Business owners want to run a business from the property they own. However, the property may be seen as a prestige generating feature of the business. Property is sometimes linked to raising finance for the business, such as:

- as security for a mortgage
- by sale and leaseback.

Most businesses have setting up requirements under an incredible range of topics. In general they include the following:

- environment — environmental law, eg the environmental impact assessment
- health and safety — fire precautions law and health and safety law
- waste — waste management law, eg landfill taxation
- pollution — the contaminated land regime

- development — planning law, eg planning permission
- taxation — see Chapters 26, 27, 28 and 29
- human resources — disabilities law, data protection law
- landlord and tenant relations — landlord and tenant law.

Leisure owners

Owner-occupation of leisure property is essentially a life-style phenomenon. The kinds of property covered under this head are not business uses. They include:

- the second home
- the allotment garden
- the beach hut or beach bungalow
- the shoot
- fishing waters
- the stable and paddock
- the fixed mobile home or parked caravan
- the timeshare.

They are properties which are almost always held away from the principal residence and used for private pleasure and enjoyment with family and friends.

Taxation

In principle, a capital gains tax liability may arise on the disposal of leisure property.

Public owners

There are nearly 200 statutes which provide for the purchase of land for public purposes. If empowered, local authorities and others may purchase either compulsorily or by agreement. (Some private sector companies have similar powers of compulsory purchase. Many were formerly in the public sector, eg companies in electricity, gas, telecommunications and water industries.)

The provision of public buildings is invariably a matter of statute under which one of the following procures the building:

- a government department
- a government agency
- an urban development corporation
- a local authority, either as a public duty, eg a school, or as a discretionary opportunity, eg a public park
- a statutory undertaker, eg Civil Aviation Authority.

In most cases where a body was set up under statute, the functions that that body has regarding property will be given in the Act, including that:

- to acquire land
- to hold land for specific purposes
- to develop land
- to raise funds
- to grant leases in land or buildings
- to sell land.

When dealing with such a body it is important to ensure that it has the power to do what is being done, ie the action to be taken, is *intra vires*.

Voluntary ownership

The ownership of heritage property is an important part of the property market in the voluntary sector. Social landlords of residential property comprise housing associations and other statutory bodies whose remit is affordable housing for those who need help with the first step on the property ladder. Finally, all manner of charitable objects are met by charitable bodies who provide accommodation for needy folk or hold property as investments (the income supporting the charity) or both.

Heritage bodies

English Heritage, the National Trust and the Landmark Trust are examples of voluntary bodies holding heritage buildings and structures for public access, leisure and recreation. Their ownership is long-term and virtually inalienable. For instance, special parliamentary procedures operate when compulsory purchase is used to acquire the property of the first two bodies.

Community interest companies

A new kind of company is now allowed, ie the community interest company (CIC), under the Companies (Audit, Investigations and Community Enterprise) Act 2004. The arrangements are such that they are expected to begin operations from July 2005. Regulations are being drafted at the time of writing (December 2004). Typical characteristics of a CIC include:

- it exists for community benefit
- it passes the test that a reasonable person should think it is for community benefit
- certain companies are excluded, eg those of political parties
- it may not be a formal charity, but charities may have CICs as subsidiaries
- it will have an asset lock, ie assets may not be distributed to its members
- some may be allowed to pay dividends to members but on a capped basis.

The advantage of CIC status is that the widely appreciated company status will embrace CICs — generally making it easier for outsiders to relate to and work with them.

Registered social landlords

Generally, registered social landlords provide housing for those who are unable to buy or rent on the open market. Many offer housing in niche markets, eg for those who are disabled. The Housing Corporation registers social landlords including:

* housing trusts
* housing associations
* companies
* cooperatives
* industrial and provident societies.

Many social landlords are charities.

Charities

Many established charities have substantial investment holdings from which they receive investment income from property and other assets. The income supports the objects the charity, being concerned with one or more of the following:

* the alleviation of poverty
* the advancement of religion
* the advancement of education
* any purpose of public benefit.

The government has indicated that the range of objects will be extended to 12 areas. If the Bill is passed, the Charity Act (2005?) will, no doubt, provide for this to happen.

Privileges

As might be expected charities hold a special place in society and have privileges including:

* freedom from income tax on investment income
* freedom from capital gains tax on realised capital gains
* bequests to a charity from an estate are not included in the value of the estate for inheritance tax (thus encouraging such gifts)
* exemptions and relief from non-domestic rates
* zero rating of qualifying charitable buildings, ie those used for charitable purposes (not those premises used for business or investment)
* on receipt of any admission fees to property, a customer who pays income tax may donate the tax to the charity.

Regulation of charities

The Charity Act 1993 regulates charities with the Charity Commission being responsible for overseeing their creation and management. Where a social landlord has charitable intentions any registration by the Housing Corporation, ie as a registered social landlord, will be accompanied by collaboration between the Housing Corporation and the Charity Commission to register the body a a charity.

(It may be noted that under the new regime for CICs, although some of them will have charitable objects but they will not be permitted to become a charity. However, charities may establish a CIC as a subsidiary.)

Finance and taxation

Charities raise funds from such sources as:

- an on-going routine, from individuals and companies by numerous events, flag days and volunteers' efforts
- by major national taxation exemptions and reliefs on their investment income and gains on their realisation of capital assets
- exemption or relief from business rates on certain properties
- exemption from capital gains tax and inheritance tax, for the donor, on gifts and bequests received
- the Balance Foundation, which seeks to obtain from the financial institutions funds which are unclaimed.

Ownership of roads

When a proposal to develop land involves the single ownership of the land on both sides of an existing road, issues arise concerning the following:

- the ownership of the land under the road
- the rights of the users of it
- the need to divert or close the road
- liability for loss or damage to users of a private road and others, eg the Occupiers Liability Acts 1957 and 1984.

The ownership of land under roads isset out in Box 6.6.

Box 6.6 Ownership of land beneath a road or a highway

Ownership by an individual, trust or
company, ie in private ownership
(not subject to rights of others)

full ownership —
- where the road is not a highway and it is on privately owned
  land upon which the public has no right of way other than,
  perhaps, with the permission of the owner of the land

As above (subject to rights of others)

full ownership —
- where it is on private land although an adjoining owner or
  occupier of land has an easement of way on the road
- the public may have limited use with the permission of
  the owner (private usage and not a highway).

Ownership by a public body (in a private
capacity) where the land was not acquired
for a highway

private ownership of the sub-soil —
- where the road is a public highway expressly or impliedly
  dedicated to the public, but the land was not acquired
  by a highway authority)

Public ownership by a highway authority

full public ownership of the substrata —
- where the land was acquired compulsorily or by agreement
  under statute by a highway authority for a highway, eg a
  motorway
- adjoining frontagers do not own the substrata

Property Information and Standards 7

Aim

To present the sources of information on property

Objectives

- **to identify roles and activities in property information**
- **to explain the principles of freedom of information**
- **to describe the nature and sources of official information**
- **to outline the kinds of standards which impact on property**

Introduction

Information about the property industry is diverse and may be obtained from many sources. This chapter looks at property information from several perspectives, namely:

- the roles and activities of information providers, generators and users
- information systems — storage and access
- where and how to obtain the information
- uses to which information may be put.

Other chapters consider information requirements in a particular context. Generally, information supports work in every aspect of the industry. It is used to effect change within organisations and in career development.

Roles and activities

Information in the property industry is generated, gathered and promulgated by a host of organisations and individuals in different guises. Box 7.1 attempts to conceptualise the roles into six generic groups, identify specific roles and briefly indicate their activities.

Box 7.1 Information from conceptual roles, specific roles and activities in the property industry

Users	clients	• to function as owners, occupiers and developers in dealing, business, investment and so on
	government (all levels)	• to monitor policies • to develop initiatives
	professionals	• to advise clients • carry out work of buying and selling, developing and managing property
	journalists and mediumists	• to interest and inform sectional interests and the public
	lobbyists and pressure groups	• to build a scenario or case • to inform
	students	• to learn • to develop a specialism
	teachers	• to inform • to develop a subject
Generators	research students	• to achieve a result or knowledge • to develop a specialist topic or interest • to obtain a qualification
	researchers	• obtain new information and insights • to meet commissioned or supported output
	opinion pollsters	• to establish held opinions • to inform
	government	• to record performance • to obtain statistics
	analysts	• for analyses • to inform
	map-makers	• to provide information regarding place, direction and an area's features
Influencers	government	• to commission • to meet needs • to analyse situations
	research commissioners	• to obtain targeted research
	professional bodies	• to meet members' information requirements • to develop a well qualified profession • to establish standards and guidance on best practice
	research foundations	• to develop a well informed society

Box 7.1 continued

Gatekeepers and controllers	government	• to enable freedom of access • to keep secrets
	clients	• to maintain confidentiality
	professionals	• to maintain competitive position • to maintain commercial confidentiality • to promote business • to manage, design, construct and operate property
	data registrar	• to protect personal data
	information commission	• to free information
Maintainers	writers	• to advance recorded or known information • to establish linkages and provide insights
	glossarians or glossary writers	• to provide insights • to inform meaning and remove confusion
Custodians and dispensers	librarians archivists information officers	• to enable access and linkages • to catalogue and store • to enable access • to dispense • to inform

---

Systems for the functions

Information in society is in transition. The historic communication continuum was from oral history — written vellum — printed text — typewritten text — word processed text — Internet text and attachments. Systems for the roles or functions indicated on the left side of Box 7.1 are in place or developing. Clusters of information are geographically separated in offices in hard copy. Examples include:

- records of title — Land Registry's local offices
- planning registers — local planning office records
- local land charges — local authority office
- flooding — the Environment Agency's risk and probabilities by postcodes
- rating list — local authority office.

Other depositories of hard copy information include the libraries of professional offices, colleges, professional bodies, trade associations and the like. However, much of the information is available in virtual form on the Internet. More importantly, the ability to access the data and manipulate it is being increasingly developed and understood. Systems for effecting transactions in chattels, shares and other intangibles already exist or almost exist on the Internet through Internet-based exchanges or markets, for example, the virtual market for emissions trading in Europe. Property searches are just about Internet ready and e-conveyancing is coming.

At another level, information is about to explode with wi-fi, ie when local authorities and others begin to appreciate the potential for street security and other community development systems.

Information freedom and law

On the one hand, much information is protected by laws intended to support authors, inventors and others. These include:

- copyright
- registered designs
- patents
- official secrets
- personal data protection
- evidence for court hearings.

Sometimes the law protects information from use during a prescribed period or an embargo time, eg company information prior to publication. Many aspects of the property industry are affected by these laws.

On the other hand, the government is attempting to lift the veil from information which needs to be in but is kept out of the public domain. However, the law also requires the provision of information, eg the Companies Act 1985 requires annual reporting and financial statements. Similarly, planning authorities may invoke discretionary powers for the supply of information as a schedule 2 environmental impact statement.

Freedom of Information Act 2000

As briefly described information is not freely available, but the Freedom of Information Act 2000 (2000 Act) is an attempt to free information which does not need to be protected by secrecy. Thus, the 2000 Act provides members of the public with the right of access to information held by public bodies. Much information is readily available and some of the public bodies providing information about the property industry are given below. It appears that about 100,000 organisations may be covered by the 2000 Act.

There are exceptions to the right of access, including:

- information which is barred by law, eg price sensitive information
- information on prospective policy or financial matters not yet determined by the public body
- information which would be expensive or unreasonable in personnel time or other costs
- personal information protected under the Data Protection Acts 1984 and 1998 concerning paper-stored and electronic-stored information respectively.

Sometimes a public body fails to supply or denies access to information. Following a direct complaint, a person who is not satisfied with the reason given, may make representations to the Information

Commissioner, who is empowered, from 1 January 2005, to determine whether the information should be made available.

Data protection

The 1984 and 1998 Data Protection Act seek to protect personal information held by employers, public bodies and other organisations. Stored information in hard copy and electronic formats must not be revealed to persons unauthorised by law to have access to it. Thus, the display to the public of CCTV and television images of seeming intruders to property may be an offence.

Insider dealing

If made public, information likely to alter the price of a quoted company's shares is price sensitive information. Any person who obtains such information from an insider and subsequently buys or sells shares in the company is almost certain to be offending under the insider dealing statutes, eg the Criminal Justice Act 1993. An insider not only includes a director or member of staff but also an outside property consultant working for the company who discloses such information, perhaps inadvertently.

Architectural plans, documents, designs and devices

The use of architectural plans, documents, design and devices without authority or within the limits allowed by law may contravene legislation designed to protect the author or inventor of such items, eg copyright, registered designs and trademarks.

Government and parliamentary information

Enactments of the government

Government policy on property which has the force of the law is primarily expressed in enactments and derived law, including:

- Acts of parliament
- statutory instruments — orders and regulations, eg the Town and Country (Use Classes) Order.

Interpretation of enactments is a matter for the courts. Many enactments include definitions of the terms used. However, the Interpretation Act 1979 provides a basis for their interpretation.

Future policy information

The government frequently publishes consultative papers on prospective policy which has not yet received statutory backing. A White Paper indicates that the policies therein are very likely to become law — either as a new area of law or as an amended law. Sometimes other coloured "papers" are used to promulgate proposed policy, including;

- Green Papers — denoting tentative proposals for policy on which views are sought
- Yellow Papers — denoting firmer policy after widespread views have been obtained (they have limited circulation).

Established policy information

Much policy is expressed as being guidance and advice. Normally, these documents should not be ignored, but they do not have the same force as an Act, regulations or statutory instrument. They are, in many instances, the government department's official interpretation of the law (but not a court's ruling). Documents, such as the following are given as examples:

- Planning Policy Guidance notes and Planning Policy Statements
- Mineral Planning Guidance notes
- Customs and Excise booklets
- Inland Revenue booklets on taxes and other matters (many cover property matters in detail).

Reports by parliamentary bodies

Numerous parliamentary bodies publish a wealth of detail on topics relevant to the property industry. For instance, the following are obtainable:

- the annual reports of the departmental agencies
- reports of the parliamentary select committees on all manner of topics, eg the environment.

European Union enactments

The legal framework of the UK incorporates European Union (EU) directives as UK government statutes. Other EU legal instruments, eg regulations, orders and opinions of the European Commission, stand in their own right. Orders and opinions, however, apply only to the party or parties to whom they are addressed.

Law — Sources

Every aspect of the law which touches upon the property industry is embedded in a range of sources, including;

- enactments — series of annotated statutes, statutes on-line (also, see the last section)
- law reports — gathered as a large number of series and published in different media, eg books, journals, newspapers and the Internet (for example www.egi.co.uk)
- law books — annotated statutes, specialised volumes (conceivably, every topic from different perspectives)
- books of forms and precedents — on procedures.

Generally, for the user who is not a lawyer or specialist professional in another field, books, law reports and so on give a picture of a topic so that knowledge, guidance and precedent may be established. Opinion may be obtained from lawyers, but the courts and tribunals will give the final decision.

Law reports

There are numerous law reports which are or have been published in series, including:

- the traditional series of reports include the general *All England Reports, Weekly Law Reports, Current Law Reports, Rating Appeals* and the *Scottish Court Land Reports,*
- the *Times Reports* — one or more short reports are published, commonly on a daily basis
- the *Estates Gazette Law Reports* come in several formats — longer reports are first published weekly in the *Estates Gazette* and then collected in volumes
- *Estates Gazette Case Summaries* are published daily on line and weekly in *Estates Gazette*
- several journals publish reports on cases or contain articles discussing cases in their context.

Audit Commission and Office for National Statistics

Under the Audit Commission Act 1998, the continuing Audit Commission is empowered to investigate the affairs of public bodies by auditing their financial reports and accounts. It also has responsibilities, for instance, to assess and review the performance of best value authorities, both absolutely and relatively to similar or peer bodies (see Chapter 23). Much information and insight may be gleaned from the commission's reports.

The Office for National Statistics (ONS) provides the nation's statistics. It is a substantial source of information into such main topics as the economy, the natural environment, population, travel, and the built environment.

Local official information

Much of the information obtained from local government, statutory bodies and other public bodies has statutory backing and stands as official policy, standards, controls and the like. The following are illustrative examples:

- the planning register
- the local land charges register
- the electoral roll
- the rating list
- price lists for services, eg electricity, to buildings
- the valuation list.

Standards

Many standards for buildings and matters which affect buildings have been formulated by various organisations. An illustrative selection is given in Box 7.1 together with other items. It illustrates the range of concerns but it is by no means comprehensive. (These and other standards are referred to in several chapters.)

Box 7.2 Standards and other matters which affect the property industry

Accounting	Company House	• company records • company accounting statements
	Public limited companies	• annual and interim reports and accounts
	REFS (Really Essential Financial Service)	• monthly analysis of quoted companies' financial affairs
Building regulation and control	local authorities (building control)	• standards, approved documents and other documents • information and guidance on the application and practices of building control in their area
	private building control office	• statistics on starts and completions • exemptions
Measurement	construction	• RICS Standard method of measurement
	buildings	• RICS Code of measuring practice
	planning	• RICS Code of measuring practice
Planning	Regime in transition as a result of the Planning and Compulsory Purchase Act 2004	• Planning Policy Guidance notes (PPGs) becoming Planning Policy Statements (PPSs), etc.
Valuation and appraisal	RICS TEGoVA	• RICS Appraisal and Valuation Manual (May 2003) • European Valuation Standards (2004)

Accounts and accounting standards

In the UK, the Companies Act 1985 lays down requirements for annual reporting and financial statements of companies. The UK Generally Accepted Accounting Principles (GAAPs) have been the basis for the Financial Reporting Standards (FRSs) to which UK accounts must conform (but see IFRS below). The standards have been established by a succession of accounting bodies — the latest being the UK Accounting Standards Board (ASB). However, now international standards are adopted.

Hitherto, some companies with operations in more than one country have been required to produce accounts to more than one set of national standards. In recent years considerable efforts have been made to establish common international standards. Of course there are GAAP standards for the USA and those for other countries, eg Australia, but progress has also been made towards global standards under the auspices of the International Accounting Standards Board (IASB) with International GAAP, ie International FRSs. Although progress has been made, from 2005 companies in EU member states are required to follow the IASBs common standards, ie International FRSs.

British standards

The British standards are reflected in almost every aspect of the property industry — environmental land management standards to standards in gardens. Many are the basis of or include aspects of European or international standards.

Planning standards

Enactments are the base standards of planning but the ODPM issues a range of documents on virtually every planning matter. Generally, the documents are advisory missives or guidance but they have a strong influence. They include:

- planning policy guidance notes (PPGs) (some have become or will become planning policy statements (PPSs))
- mineral planning guidance
- circulars — now tending to cover procedures, but may include standards.

Local planning authorities issue standards on many topics, eg housing density, car parking, materials. The standards are usually based on local situations and concerns, but reflect national policy and guidance.

Valuation standards

For property valuations which are required for company reports and annual accounts, the valuer works to a specified set of standards, including those of the following organisations:

- the International Valuation Standards Committee — International Valuation Standards (IVS 2003 ed)
- the RICS — Appraisal and Valuation Standards 5th ed 2003
- TEGoVA — European Valuation Standards 2003 ed.

The client may specify standards to which the valuer works, eg those of the USA. The RICS standards contain both mandatory requirements and discretionary matters.

Information services and other sources

The role of professional bodies in the information function of their work is outlined in Box 9.1 in Chapter 9. Much information is in the public domain. Box 7.3 shows a few of the many kinds of topics on which data is available, their sources and the kind of information.

Other sources are, of course, the daily newspapers, monitored by EGi, and the many weekly journals, research series, magazines of professional bodies, trade associations and professional firms and the like.

Box 7.3 Some of the information services available to those in the property industry

Building costs	Royal Institution of Chartered Surveyors	• information on the unit costs of most types of construction
Building maintenance costs	Royal Institution of Chartered Surveyors	• information on the unit cost of most building maintenance services
Business rates	Valuation Office Agency	• information on rateable values published in rating lists for local use
		• programmes for appeals after a revaluation
		• information leaflets and guidance on practice
Council tax	Valuation Office Agency	• information on bands for council tax
		• information and guidance on council tax practice
Electoral roll	local authorities	• electors by name and address
Environment	Environment Agency	• air pollution
		• flooding database by postcode
		• radon
Local land charges	local authorities	• data on individual properties
Land registration	Land Registry	• records of title to land and buildings, eg freeholds and leases (over seven years duration)
Rents	rent officer	• old and current records of fair rents
National taxes	Inland Revenue	• information and statistics about taxes
		• IR information and guidance on taxes
		• practice statements

Model documents and codes for contracts

Once the parties have agreed to a transaction, a formal contract is usually prepared for their signatures. (This is not invariable, eg where there is verbal agreement.) Common forms or model documents are available for use in given situations, saving time by not starting from scratch every time.

Numerous model documents exist as good practice for contracts, agreements and licences in several fields of the property industry, including:

- the ICE Conditions of Contract
- the JCT forms of contract
- the statutory model clauses of the agricultural tenancy
- BPF agreements for developers and archaeologists
- the RICS's model farm business tenancy
- the RICS's model grazing licence.

In general the documents are drawn up by professional or official bodies with the view to establishing a benchmark document. In drafting a contract the benchmark stands, parts of it stand or it is amended by the parties to suit the requirements of the situation. In the case of the model clauses for agricultural tenancies, silence on a matter in an established agreement will result in the application of the relevant model clause.

A similar purpose is intended for the voluntary Codes of Practice for Commercial Leases prepared by the British Property Federation and other bodies in the property industry. (Aspects of contracts for leases are given in Chapter 3 and building contracts are included in Chapter 19.)

Land registration

There is a national Land Register maintained by the Land Registry for England and Wales. (Scotland has is own separate system of registration.) One of 26 local offices of the Land Registry maintains the part of the register for its area. There is a requirement for compulsory registration under the Land Registration Act 2002 in the kinds of transaction listed below:

- sales of freehold and leaseholds
- grants of certain leases
- transfers of land under a will or intestacy.

Owners of unregistered property may register it voluntarily. This is desirable in the following instances:

- a sale or other transaction is proposed in the future and time may be important
- precise boundaries need to be established prior to a disposal.

Statistics

The Land Registry publishes a quarterly Residential Property Price Report of actual unadjusted average residential sale prices recorded in England and Wales at the date of completion. Allowing for the duration of conveyancing the prices will lag the date of the negotiated prices. (The web-site is www.landreg.gov.uk.)

Planning and building control information

Local authority offices hold much information of value to the property industry including, in particular, the planning office and the building control office.

Planning register

The planning register gives information on the planning history for each individual property (see Chapter 16 for the site appraisal context).

Local Land Charges Register and Public Information Records

The Local Land Charges Register and Public Information Records contain information about the properties in an area. A buyer or developer of a property may find important background information. Such searches are a prerequisite to the purchase of a property and are normally the responsibility of the buyer's solicitor.

A Land Charge Search for information on a property may be made in person for a small fee at the local offices of the local authority. The applicant must:

- complete a form LLC 1 for the Certificate of Search
- complete a CON, being an Enquiry Form
- supply a plan showing the property and its locality.

The information which may be obtained includes:

- compulsory purchase orders
- conditional planning permission
- drainage
- enforcement notices
- financial charges
- highway proposals
- legal restrictions
- noise abatement
- tree preservation orders
- smoke controls in the area.

Building control

The local authority, eg the district council, maintains a building control office responsible for the approval and supervision of construction in their area (see Chapter 14 for more information). Apart from regular information about the building regulation, the office may hold special information, eg local radon risks and flooding problems. They also have statistics about building completions.

Values, prices and building costs

Information and analyses of land and property prices and values is supplied by a number of organisations, including:

- the Land Registry
- rent officers
- the Valuation Office Agency
- building society research offices.

Of course local conditions and site conditions will affect the value of a particular site — access, contamination, the need for demolition and so on.

Building costs and building maintenance costs are available from the RICS, ie the building cost information service (BCIS) and the building services costs information service. Several books showing prices are also available. Building costs vary in different parts of the country. The user should expect to be able to vary the standard figures with an adjustment ratio. Similarly, local conditions may affect the price — demand, traffic congestion, lack of on-site storage and so on.

Markets in Property

8

Aim

To explore the nature of the sectors of the property market

Objective

- to briefly describe aspects of the principal sectors of the property market, eg agriculture, forestry and retail

Introduction

Although the amount is barely comprehensible, the value of property in the UK market is of the order of £3 trillion. This huge market is generally regarded as being divided into broad sectors each comprising similar property, eg agricultural property. Within each sector, subdivisions may be discerned but different classifiers may take the properties in the same sector and use and name them differently to suit their purposes.

This chapter takes the property market's sectors and somewhat arbitrarily describes them in terms of general subdivisions. An overall picture of the trends, concerns and issues of each sector will be included.

Agriculture

The rural economy supports a range of land uses and different kinds of property within each. Broadly, the following is a flavour of the total:

- farming — the dominant use which itself may be categorised, eg arable
- horticulture, market gardening, and nurseries
- forestry, woodlands and short rotation coppicing (see the next section)
- retailing — garden centres and farm shops
- stately houses and historic homes, parks and gardens

- sporting — fishing, hunting, clay pigeon shooting
- leisure — water-based sports, cross country, walking centres, public houses.

Each land use is a potential market but in this section farming is considered.

Farms, estates and houses

For statutory purposes agriculture covers diverse properties from farming estates to some kinds of allotment. In this chapter the treatment is limited to farming. Occupiers of farms are mainly either owners or tenants: but some are research institutes or charitable trusts, such as Brogdale Trust in Kent, which has the national apple tree collection.

The large farming estates are tenanted but an owner-occupied or managed "home" farm is common. Other property comprising part of an estate might include:

- sporting rights — enjoyed by the owner's family and friends, let out or ran as an estate business
- mineral rights — held dormant or worked by the owner, let out or reserved to another person
- stately home or country house — open for public recreational access, conference centre or other business opportunities.

The high cost of upkeep of stately homes and the conditions for exemptions and reliefs of agricultural property as a favoured asset for inheritance tax, make the business of house and estate opening very common.

Tenures

Traditionally, a farm is owner occupied or held on a lease or an annual tenancy with succession to the farmer's family. Legislation controls the tenancy in the following ways:

- security of tenure — the farm is held for life and there is one succession
- rent — the rent is revised every three years to the open market rent with disregards for:

    (a) tenant's improvements
    (b) the fact that the tenant is tenant

- compensation — when a tenant gives up possession, compensation may be payable by the landlord for the likes of

    (a) tenant right (the tenant's maintenance of the soil)
    (b) tenant's improvement
    (c) loss of possession.

Farm business tenancy

A new form of tenure was established under the Agricultural Tenancies Act 1995 called the farm business tenancy. It copes with the increasingly common diversification in agriculture. Farms are endeavouring to broaden the range and smoothness of incomes from the land, thus coping with the fluctuations of income from traditional farming.

Diversity

The trend of diversification has resulted in some novel new businesses being allowed in redundant farm buildings, including:

- mazes and other labyrinths
- gift shops
- children's events
- industrial workshop uses
- retailing of dried flowers and homeopathic remedies.

Investors

Agriculture is a sector which has a following of investors, including:

- the Crown Estate and the Duchies of Cornwall and Lancaster
- the Church of England
- the National Trust and other charitable foundations
- the financial institutions — pension funds, insurance companies and the like
- property investment companies, and property investment trusts.

Grants, taxation and finance

Generally, the decline of parts of the agricultural sector and the increasing awareness of the environment has seen a shift of government policy. In the main less support is available for food production but increased support goes to the natural environment. This is reflected in the grant schemes where the farmer has a range of grants and financial incentives available, including:

- apple orchard grubbing, ie removing apple trees
- set-aside financing, ie releasing land from production
- farm woodland premium scheme to assist diversification
- payments for nature conservation, the maintenance agreements payments.

Taxation, including valuations for taxation is dealt with in Chapters 26 to 30. Each grant and payment under the schemes mentioned above need to be looked at do determine whether they are taxable.

Forestry

Box 8.1 shows the principal roles and activities of many in the forestry industry.

Forestry is an ancient traditional use of land going back to beyond the time of William the Conqueror. In modern times the principal purpose is in owning a forest is the growing and harvesting of timber. Direct private investment is made by purchasing woodlands or by purchasing land and undertaking investment in afforestation, perhaps with a grant from the Forestry Commission.

Shares in a forestry company or investment property trust allow indirect investment to be made by individuals and others.

Box 8.1 Forestry industry

Owners	• farmers • forest and woodland timber growers • Crown Estate • National Forest Ltd • Woodland Trust and other conservationists	Not all owners will have these: • owns land for growing and harvesting • owns land for public recreational access • owns land for conservation
Manager	• individual directly or indirectly employed to manage operations	
Land agent	• individuals, companies or consultancies involved in forestry management on behalf of an owner	
Forestry Commission	• owns land for growing and harvesting for timber • has a policy of public recreational access • manages the woodland grant system • issues licences to fell • advises government on forestry policy	
Contractors and subcontractors	• contractors are employed by the owner to plant, harvest, or build roads and structures • subcontractors employed by contractors, perhaps as specialists	
Buyers	• those who buy timber, cut or as standing trees, from landowners	
Professional bodies	• Forestry Contractors' Association, RICS	

Although timbers from mature hardwood trees and conifers are the principal output of the industry, other products and pursuits include:

• coppice from immature trees
• charcoal
• rural style furniture
• wooden plates and other turned items
• traditional forms of fencing, hurdles, poles and posts
• hunting
• green burial;
• leisure and recreation, eg war games
• wildlife conservation.

Investors

Investors in forestry include:

• the Forestry Commission, but it has sold much of its strategic forestry land
• quoted forestry investment companies and property investment trusts
• the institutions, but limited perhaps
• individuals — perhaps as a favoured tax avoidance shelter.

Taxation

Inheritance tax relief makes forests and woodlands a favoured asset, subject to the observance of conditions about ownership and felling. Provided there is no sale of the timber the relief may be enjoyed through the generations of the original owner's family.

Generally, woodlands are a substantial source of notional total return in that they are sheltered from income tax and capital gains tax. Of course, those who work the woodlands, rather than merely hold them, pay income tax on profits, including:

- growing Christmas trees
- growing short rotation coppicing.

Forestry Commission

For many years the Forestry Commission was the principal landowner and producer of timber on behalf of the government who saw forests as part of the national defence strategy. Its role is advisory and it administers the grant awards scheme for private sector investment in new plantations. It also controls felling by issuing of licences.

From the 1970s there has been an increased expenditure on facilities for access by the public and on leisure and recreation facilities. Since the 1980s the nation's forests have been sold, so reducing the holding in the public sector. As a result, private and voluntary involvement in forestry and woodland management has increased.

More recently, with the decline of traditional agriculture, woodlands have again become a policy thrust, with the creation of the first National Forest in the Midlands.

Industrial property

Much of the heavy industry has gone and lighter manufacturing, technology based industries have come about. Developments reflect the change with hybrid uses which contain research, light manufacture and ancillary services. The quantum of each in a property may vary as time passes.

Many of the old mills and premises of the old industries are now derelict with resources being used to restore the land to developable quality. Urban regeneration is the theme for much of the construction work in progress. The urban development corporations are at the forefront of large area regeneration, eg Thames Gateway. Similarly, English Partnership is the largest non-government public body involved in regeneration.

Investors

Investment in industrial property is broadly the same as for agricultural property. There are, however, investment trusts and private investment funds which specialise in investment in quoted and unquoted manufacturing companies and service companies which will hold property for business operations.

Taxation

Businesses are taxed under schedule D case I on the profits of the manufacturing or other industrial activity. Capital allowances are available and must be claimed for qualifying buildings.

Leisure property

The leisure industry has a wide range of property market subsectors, including:

- heritage property tourism
- hotels and hospitality property, eg public houses
- cinemas, theatres and concert halls
- sports stadia and grounds
- racing, eg horse racing and dog racing
- water-based sports, eg swimming, surfing and sailing
- leisure centres and clubs
- golf courses.

Leisure development tends to come about in phases. Marinas are developed, followed by ice rinks and equestrian centres; only to fade as a new mood takes up a new development. At the end of 2004, gambling reform seems to be generating a phase of potential casino developments.

A number of sports and other activities depend on one or more technologies. A typical stadium will embrace, among others:

- audio-visual systems for information and replay as well as crowd control
- surface technology for grass sports or water cum ice sports, etc
- crowd movements and refreshment facilities
- cleaning and waste management
- weather control roofing systems
- timing and score keeping
- conference facilities, including audio-visual aids, network and internet technologies and dining facilities.

The technologies are listed but the marketing mixes which could be offered will range from those required for an Olympic event to a pop concert.

Casinos

In the future, the hesitantly forthcoming Gambling Bill (2004), if passed, changes the way in which casino development is controlled. Until now, "permitted areas" were localities where a casino could be established, subject to:

- obtaining planning permission from the local planning authority
- obtaining approval to the proprietorship
- obtaining a licence from the Gaming Board
- gaining approval to members of staff from the Gaming Board.

Casinos were for members only and could not offer live entertainment. Finally, the number of gaming machines was limited.

The Bill removes the designation of permitted areas and substantially liberalises the casino industry. As many as 100 proposals for casinos have been or are being prepared for submission to the planning authorities in anticipation of royal assent. However, government policy has been evolving so the arrangements are not yet certain, eg there may be far fewer casinos than expected, etc.

Forest-based leisure

From the 1970s the Forestry Commission woodlands have been opened much more readily for public recreational access. This was a deliberate policy of government at the time. Walking and orienteering are common activities but holiday accommodation is offered at some places.

Walking

Walking is probably the most popular past-time for the non-specialist member of the public. The Ramblers Association and the National Trust and other sight-seeing organisations generate millions of visits to rural areas. The right to roam in open countryside was introduced by the Countryside and Rights of Way Act 2000. It does not cover coastal areas but the Act allows for extension of the right to roam. The right may be regarded as a fillip for the economic development of the opened areas, as more of the population are attracted to walking.

Water-based leisure

Like the forestry industry, water has been opened up to public recreational access. Here, there may be health issues which prevent the full scope of recreational usage on reservoirs, eg motorised boating.

In some seaside areas, eg Newquay and Woolacombe, surfing the waves has developed into a substantial tourist economy. At Bournemouth consideration has been given to creating an artificial reef with motor tyres — so as to generate waves for the surfers.

Mines and quarries

Mineral rights

Generally, mineral rights vest in the owner of the land, except coal, gold, gas, oil and certain other minerals are subject to the Crown's rights. Box 1.4 in Chapter 1 summarises the position on such minerals and notes the official bodies involved for each mineral. Where the minerals are on the foreshore or on the seabed the section on the Sea at the end of Chapter 1 briefly covers the position.

Planning

Generally, mines and quarries, particularly opencast, ie in essence, mining on the surface have the following planning attributes:

- apart from planning permission many proposals will require an environmental impact assessment
- some sites may be suitable for landfill but care must be taken to ensure that the water table and air in the locality is not polluted
- aftercare and restoration conditions are likely to be required to get any damaged land back to agriculture or some form of development.

Under the planning regime introduced by Planning and Compulsory Purchase Act 2004, county planning authorities retain their role as mineral planning authorities. Planning policy on minerals in provided by the mineral planning guidance notes (MPGs).

Landfill tax

A conditional exemption from landfill tax is available to certain mineral waste arising during commercial operations and processing at mines and quarries. Thus a quarry with a conditional planning for restoration by refilling wholly or in part would enjoy the exemption. This applied from 30 September1999.

Offices

Offices range from a room in the owner's residence to a multi-storey prestige headquarters building. Also, from a functional perspective one could include offices which are ancillary to other uses, such as an industrial property with offices in the building.

Although many offices are owner occupied very many more are leased to individual tenants or to several tenants in one building, ie multi-let property. The former tend to be held by the tenants on full repairing and insuring terms: the latter on internal repairing terms.

Typically various patterns of layout are experienced — from individual offices to open plan. Modern thinking on design has moved to include atria, rest areas, meeting rooms and snack bars.

Hot-desking

In some businesses, particularly where staff are part-time or full-time but spend some of their working time away from the office, shared work stations are provided and are becoming increasingly popular in reducing the cost of accommodation.

Serviced offices

An innovative use of premises is the serviced office where a range of business support services are offered to largely short term users who may want to hire the premises for part of a day to, say, several months.

Home working

With the growth of information and communications technology (ICT), working from home or even abroad has become feasible for many who would otherwise go to the employer's office.

Call centres

A call centre is a telephone-based operation where ranks of cubicles house staff whose job is to answer enquiries from customers or those seeking information. An employer may operate a call centre within the organisation or the work may be outsourced, even abroad.

Shared service centres

Some international companies with business operations in several countries find it worthwhile to take the staff of a business function, eg accountancy, and move them to offices in one location. Unlike some call centres the work is outsourced by the employer.

Investors

Again, although stronger, the investment market has a similar range of players to that for agriculture, excluding any specialised investors in agriculture. Indirect investment is enjoyed by individuals and others owning shares in property companies.

Residential property

Trends

Several trends may be discerned in the residential sector. Box 8.2 identifies examples and briefly describes them.

Box 8.2 Trends in the residential property market

Private sector provision	• growth in the number of households, estimated at 189,000 a year • long-term trend to private owner-occupation • growth in life-style provision, eg gated settlements, marina-side and golf-side settlements • growth in the ownership of the second homes
Public sector provision	• decline in local authority role as a provider • growth in right to buy take-up • issues of housing, services and opportunities in rural communities
Voluntary sector provision	• growth in registered social landlords' role in place of local authorities • greater segmentation in provision, eg key worker and individuals with disabilities • availability of different offerings, eg rentals, shared ownership and right to buy • long-term growth in provision for the care of the elderly

---

Housing and regeneration

Many industrial, residential or mixed areas in cities and towns have suffered decline for many years. As far as housing has been involved, various means have been attempted to halt or reverse the process, including:

- estate development — with prefabricated houses after the Second World War
- general development areas — for bomb damaged places after the Second World War
- new towns programme — to move population to relatively well designed settlements
- slum clearance — with clearance areas, closing order and demolition orders (Part III of the Housing Act 1957)
- London's out-county estates and "housing gain" sites (Part V of the 1957 Act)
- London's decanting of elderly council tenants to seaside bungalows
- regeneration — with improvement areas and various kinds of housing grants
- growth of involvement by social landlords in improving and letting dwellings
- shared purchase by lessees and social landlords
- self-build housing developments
- more recently, supermarket developers are building house as a planning obligation for supermarket planning permission.

Life-style residential developments

Changes in life-style are inducing reflected changes in the provision of modern accommodation. New styles of development include:

- gated developments — where the security of an estate is high with serviced entrance gateways, CCTV, guard patrols and centralised alarm systems
- leisure developments — where the development is built on one or more leisure activities — golf fairway developments or marina-side development
- condominiums — studio flats, flats, apartments and maisonettes built over or around a leisure centre, gymnasium, pool, restaurant and bar
- student accommodation — furnished studio or small flats with appropriate information and communication systems, storage and parking.

Flats over shops and commercial property

In an endeavour to encourage the reuse of vacant accommodation over shops and certain other property, the government has introduced capital allowances against profits for income taxation. Conditions must be observed in preparing a proposal for a conversion to flats.

Registered social landlords

The registered social landlords (RSLs) include housing associations and certain provident societies. Development by these voluntary sector landlords has declined in recent years and is now expected to be between 19,000 and 20,000 in 2004–2005. In 2003 it was down at nearly 14,000 social homes.

However, the Budget of March 2004 showed that £20bn of government funds are planned over the next three years, with a target of nearly 30,000 for 2007–2008.

Retail property

Retail premises is a loose term for property that shoppers go into to buy goods which are displayed. It might include covered markets with lock-up kiosks. Generally, it ranges from kiosks and corner shops to large departmental stores, retail warehouses shopping centres, designer outlets and the like. More precisely the Use Classes Order gives Class as A1 use for sale, display or service to visiting members of the public. Shopping centres also have somewhat similar premises which sometimes house estate agents, banks, building societies, betting shops and the like, but these are Class A2. Finally, there are other use classes which could use similar premises, eg use for the sale of food. In this section no clear division is needed.

The popularity and growth of the motor car has changed shopping patterns markedly in the last 30 years with regional shopping centres, such as Bluewater in Kent becoming common. Other trends in shopping are indicated in Box 8.3.

Box 8.3 Trends in shopping

Consumer preferences	• growth of the use of the motor vehicle
	• growth in consumer credit
	• higher proportion of disposable income in the family budget
	• foreign travel leading to changing retail preferences, particular food
	• technology leading to growth of electronic consumerism
	• consumer consciousness of value for money, willingness to bargain and to complain
Provider responses	• consolidation of the major retail companies, either for merging like businesses or for amalgamating or sub-consolidation of retail divisions
	• provision of different retail formats, eg out-of-town shopping, retail warehouses, outlet centres, and petrol filling station retail sections
	• provision of credit facilities
Out-of-town shopping centre	• the motor car has driven the growth of out-of-town shopping centres, retail warehousing and outlet shopping
retail warehouses outlet shopping	• technologies have altered modern distribution systems — based on technology controlled warehousing and telematic and telecommunication technologies
Corner shops	• a long term decline in the corner shop may be changing
	• several 'names' have developed to share buying power, promotion and other activities
	• some large companies have entered the local shop sector to develop formats under their brands
	• several petroleum companies are providing retail sections at their petrol filling stations (PFSs)
	• some retail companies have procured PFSs and added retail sections
Petrol filling station outlets	• several company operators have in recent years provided retail sections for food and other corner shop style formats
Energy innovation	• new energies are leading to new types of refuelling stations, eg LNG

Tenures and sale and leaseback

Almost all shops are held as investments and leased on occupation leases (see Chapter 3). The investors include the financial institutions and property investment companies.

Some retailers prefer to hold their shops on a freehold basis but less so since the sale and leaseback era of the 1960s (when retail companies were taken over and the purchasers sold and took the premises back immediately on a leasehold basis). Nowadays, companies themselves tend to realise their freehold assets at market value and continue trading as leaseholders.

Storage and distribution

Changes in technologies account for many innovations in the distribution and storage property sectors which generally include warehouses, distribution centres and self-storage.

Distribution centres and warehouses

Logistics concerns the movement of goods from a supplier to customers (the customers may be external retailers (or others) or the supplier's own outlets. Technology in distribution centres seeks the following storage and distribution functions:

- receive goods
- issue receipts
- store the goods
- guard the goods
- batch or otherwise configure them for delivery
- load the goods
- deliver them to the customers
- invoice the customers
- schedule the deliveries and monitor them
- deal with any complaints of damage or non-delivery of goods.

Although the steps are conveniently described the technology can be complex, eg the scheduling of the deliveries and the delivering of goods may use telematics technology — a relatively new technological development.

Self-storage units

Self-storage has developed in the UK in the last 10 years — in the USA self-storage has been common for many years. There are now at least four quoted companies on the Alternative Investment Market (AIM) with consolidation having taken place over the last year. A marketing mix for a self-storage facility is given in Chapter 5, Box 5.1.

Container ports and dockside facilities

The decline of the docks in the Port of London has been accompanied by container ports in the Thames

Estuary. A similar trend has been experienced elsewhere. As a result many of the traditional ports have been regenerated as tourist centres and residential settlements, eg Cardiff and Chatham.

Other water-based facilities have been constructed as ferry ports, oil or gas ports and wharfs for dredged minerals.

Waste disposal

The waste industry is governed by a strict waste management regime run by the Environment Agency. Also, under the Planning and Compulsory Purchase Act 2004 waste management (and minerals) are under the retained jurisdiction of the county councils for planning purposes. Planning for waste is developing rapidly with many new forms of waste disposal have been created in the last 30 years or so, including:

- more highly regulated landfill sites
- incineration of waste
- recycling and processing of waste to create other products.

Landfill sites

Landfill is still prevalent but new EU requirements mean a stricter regime on the location of sites for landfill. Broadly, waste is categorised into household municipal waste and is inert. Some types of industrial waste are "active" and require special landfill treatment. Box 8.4 gives a description of the various kinds of waste (some licence has been taken to extend the concept beyond the usual classifications).

Box 8.4 Types of waste and their management

Clinical waste	• generally, collected and incinerated by specialist contractors
Dental waste	• see clinical waste
Gases 'waste'	• generally, released to the atmospheres after basic treatment to cleaning • coal fired power stations may soon have the facility to clean the gases in the flues, so extending their write-off availability • heat transfer and combined heat and power applications may be possible
Grey water "waste"	• generally, run off for treatment by the water companies • some companies recycle their water for industrial or cooling processes • interest is developing recycling facilities within residences
Hazardous waste	• special storage, treatment and removal required
Household waste	• see municipal waste
Industrial waste	• see municipal waste • may require special treatment

---

Box 8.4 continued

Medical waste	• see clinical waste
Mineral waste	• recycled and used in restoration or creation of landscape, eg spoil from the Channel Tunnel was used to create Samphire Hoe on the road from Dover • problems of stability of old slag heaps
Municipal waste	• collected by the waste collection authorities or their contractors • disposed of to recycling centres, landfill or incineration
Nuclear waste	• civil nuclear legacy from the nuclear power stations is the remit of the Nuclear Decommissioning Authority • empowered to clean up sites and safely move hazardous materials
Nursing waste	• see clinical waste
Pharmaceutical waste	• see clinical waste
Radioactive waste	• four levels of such waste • special treatment required for storage at a safe site • Committee of Radioactive Waste Management is to make recommendations
Rainwater 'waste'	• collected by occupiers for on-site treatment and use or for gardening • run off to municipal or water company treatment centres • problems of flooding etc.
Smoke 'waste'	• prevention or control by smoke control legislation • treatment on-site • release to the atmosphere (see natural rights in Chapter 4)
Sewage	• run off to municipal and water company treatment plants or collected from cess pits • development of household (remote homes) of on-site biological treatment eg by natural treatment and reed bed water recovery methodology
Street cleaning waste	• see municipal waste
Veterinary waste	• see clinical waste

Part 3

Professional and Contractual Services

Professional Bodies and their Members

Aim

To identify and describe the professional bodies involved in the property industry

Objectives

- **to describe the organisational structure and activities of a schematic professional body**
- **to outline the education, training and development of a typical professional**
- **to briefly describe the principal groupings of professionals**

Introduction

The UK property market includes an ancillary market of professional services with individuals who, for the most part, belong to one of more professional bodies and sometimes a niche speciality group within a professional body.

A professional body has several functions, namely:

- as a builder of a professional ethos and morals for the protection of the public, clients or customers and members
- as a learned body — being a wealth hoarder of professional history and knowledge which leads them to the generate research and contribute to education
- as creator of standards of professional and technical best practice, leading them to adopt, seek out or devise better methodology and apparatus.

The property industry is directly and indirectly well endowed with professionals; either as property professionals, eg architects, conveyancers, and surveyors (of different types), or as professionals in other fields eg accountancy and lawyers, who specialise in property matters. Brief details of the principal professionals are given below. However, many of the other chapters contain more details of the roles and the activities undertaken in various contexts of the UK property world.

Roles and activities

The roles and activities dealt with here are rather different to those in other chapters since they concern the components of the structure a professional body rather than a professional or other person. Most professional bodies are concerned with the matters shown in Box 9.1.

Box 9.1 A schematic professional body — principal functions and activities

Membership	• grades of membership
	• recruitment and induction of members
	• imbue members with professional standards
Education	• liaise with academia at all levels
	• promote academic and professional courses at appropriate levels
Training	• ensure members' early education is updated and extended, eg with continuing professional development
	• identify emerging training needs and meet them
Research	• identify research needs and priorities
	• identify academic and technical research requirements
	• encourage members and others to research
Information	• store and preserve professional information
	• give members, academics, students and others access to information
	• publish research and technical data
Standards of a technical nature	• formulate and publish technical standards of practice — mandatory and discretionary
	• liaise with relevant bodies outside on standards
Professional conduct	• formulate standards for professional conduct
	• publish and promote them to members and others, eg clients, government
	• cover such matters as: professional indemnity insurance (PII), clients' accounts, procedures for complaints against members and unbecoming professional conduct
Discipline of members	• set up disciplinary procedures
	• enforce the standards members are required to adopt
Promotion of the profession	• promote the profession to potential clients and other stakeholders, including the membership

---

Professional bodies

Unlike continental Europe which has somewhat consolidated professional groupings, the United Kingdom has many kinds of professionals who work in all three sectors, ie the private, public and voluntary sectors of the UK property world. Although a degree may be sufficient in terms of knowledge, most professionals gain credibility, and hence acceptability with clients and employers by being a member of a professional body.

Although it has many very important functions, a professional body's principal role is developing its members' knowledge and capabilities by initial education and subsequent training; other functions support and extend this role for and, in effect, on behalf of the private or public client. The functions of a typical professional body are outlined in Box 9.1 above.

Property professionals

Traditionally, the word professional is either a noun or an adjective to indicate someone with a specific status with many constructs. In some ways, in modern times, it has become a cliché to denote any one who wishes to be so described. In the property industry the term is used to denote membership of a professional body which was formed in the past with the perspective, in part at least, to protect clients and the public at large from the misdoings of practitioners. Nowadays, membership usually requires evidence of education or experiential learning, qualification to undertake the work proclaimed, continuous development and the carrying of professional indemnity insurance. Individual bodies have different requirements but those given are typical.

A particular body may have more than one grade of membership, eg fellow, associate member, technical member, probationer and student or pupil. Generally, much more detail of their work is given in other chapters. Similarly, a profession may have more than one kind of professional. For instance, in the law, a lawyer is either a legal practitioner who is a barrister (who may be a Queen's Counsel (QC)) or a solicitor.

The main property professionals are described below, together with their principal professional bodies and a brief note of the kind of tasks they undertake.

Although most property professionals would have come to the property industry and began to work in the private or public sectors in their chosen field, many have strayed into cognate fields. Thus, some have developed as specialists in judicial roles, eg as arbitrators, or academic roles, as researchers, etc. Others are not advisers to clients but are clients in a diverse range of companies and other organisations.

Education, training and development

Each profession has its own routes to full practising status. Within a profession there may be more than one professional body, each with such requirements as follows:

- levels of entry
- pre-entry levels of educational attainment
- pre-practising attainments
- levels of membership
- continuing professional development standards
- certification to practice standards.

Higher education

Entry with a university first degree in a cognate subject is usual for most professions. Entrants with a non-cognate degree may enter some professions after or while taking a conversion course at post-graduate level. Full-time education at degree level is the most common mode of entry, but part-time education suits many would-be professionals.

Part-time education may be by degree or some other acceptable mode of education, eg a professional body's own course and examination. Having obtained a first level degree and after some professional experience, many go on to master or diploma level study in a specialist cognate subject. Others take a masters programme in business studies or carry out research for a master or a doctorate.

Probationer or trainee status

The steps to full membership of a professional body vary considerably between bodies. Typically, once an individual has completed a full-time cognate educational programme, probationer or trainee membership is usually available to the aspiring professional. Full membership will normally follow once a professional experiential programme has been completed to the satisfaction of the professional body. Such a programme might include:

- structured work experience
- a logbook which must be submitted to the body
- a written portfolio of work achieved
- an interview to discuss work done and knowledge of the profession.

Continuing professional development

Most professionals are required to undertake post-qualification training of at least, say, 30 hours a year. It is sometimes known as continuing professional development (CPD) and will cover the following:

- relevant, general studies or training to refresh, update, extend or maintain professional knowledge of everyday practice
- short-term specific studies or training to develop appropriately as a specialist in a new field — a common feature of developing for a niche market
- post-graduate study in the field or a field which is synergetic for a major personal development.

The above are not concerned with developing the individual in professional leadership or management but in professional development. Of course some knowledge and skills may be transferable.

Roles

The principal role of the professional is, of course, that of adviser or expert practicing a profession directly to individual or to corporate or other clients in one of the property sectors. However, other less direct roles are within one of the following:

- a judicial or adjudicating function, eg an arbitrator, adjudicator and member of the Lands Tribunal
- business manager or administration function
- an education or training function, eg a teacher
- a research function, eg a researcher
- information or advisory function, eg a journalist
- evidential function, eg an independent expert witness.

Architects and designers

Those who design buildings are either architects or designers. The professional architect is officially regulated by the Architects' Registration Council of The United Kingdom (ARCUK). A person may not, therefore, practice under the title of architect unless so registered. As well as architects, others may offer building design services for buildings, extensions or other structures, eg a chartered building surveyor, but not as an architect.

The principal professional body is the Royal Institute of British Architects (RIBA) a self-governing body under a Royal Charter. Membership is open to individuals.

Firms of architects range from a sole practitioner to substantial international practices of several hundred persons.

Builders

Routes to professional status for individuals in building include membership of the Chartered Institute of Building or the Royal Institution of Chartered Surveyors. The latter has grouped its members into 16 faculties. Several faculties are building and construction, including:

- chartered building surveyors
- chartered quantity surveyors.

Conveyancers

Traditionally, the role of conveyancer has been and still is practised by the solicitor but for several yearsindividuals who are not solicitors have been licensed to convey land and buildings.

Estate agents

Professional estate agents

Estate agents tend to specialise in a particular property sector, such as industrial property, residential property or commercial property. For all sectors a prudent prospective client engages an agent who is a member of a professional body or is with a firm whose partners or directors are members. For corporate clients who are buying and selling property this is generally the rule of thumb. In the residential sector knowledgeable clients selling a house or flat will normally engage a professional or a professional firm.

In such instances the client has a relatively high degree of protection against breach of contract, negligence, misconduct, fraud or other criminal behaviour. Box 9.1 indicates typical safeguards that a client may expect, eg voluntary membership of the Ombudsman for Estate Agency (OEA) scheme.

Other estate agents

At present anyone may set up office as an estate agent. Such a person may have little or no experience and little or no training. There is no requirement to belong to a professional body or to join the OEA scheme. Also, there is no specific government control, such as licensing.

Building surveyors

Professional building surveying work covers a range of activities, including:

- structural and other surveys of buildings
- preparation of repair and maintenance specifications and schedules
- design of extensions and additions to buildings
- project management of building works, including snagging
- preparation of insurance claims for loss or damage to buildings following an event.

The principal professional body for building surveyors is the Royal Institution of Chartered Surveyors, which has a faculty devoted to the specialism.

Geomatics or land surveyors

Professional geomatics surveyors are represented by the Royal Institution of Chartered Surveyors which has the faculty of geomatics concerned, in particular, with that area of work, eg land surveying or hydrographic surveying. Professional engineers in certain disciplines also practice land surveying, particularly in civil engineering. Similarly, builders set out buildings and measure construction works.

Accountants

Accountants in the property world offer, of course, the usual business services of accounting for financial statements. However, taxation matters are particularly important to owners, occupiers, contractors and others. There are several professional bodies for accounting. At the end of 2005 five of them were considering a merger into a single organisation.

Lawyers — barristers and solicitors

Barristers

As lawyers, barristers, in particular, have an advocacy role in the courts and a specialised legal advisory role on complex legal matters, generally as a Queen's Council or "silk". They are all members of the Bar. Often barristers work in business, government and other occupations where their knowledge and skills may be needed. Many barristers began their professional life as surveyors, construction managers or in other fields and took up law at a later stage.

Solicitors

For many years all solicitors have had rights of audience to the county court which has unlimited jurisdiction. In recent years about 2000 solicitors have had rights of audience in the higher courts (in an advocacy role). However, a solicitor's principal role is to proffer legal services, including in such property areas as:

- conveyancing of land and buildings
- probate and the administration of estates
- property and matrimonial divorce
- setting up entities for business, including property investment and dealing.

Also, many solicitors are directors, partners or staff in companies or partnerships. Others are employed in government, education and training, business and not for profit organisations.

All solicitors are members of the Law Society, which is a self regulating professional body.

Letting and managing agents

The position of letting and managing agents is similar to that of estate agents, except that there is no independent ombudsman scheme. However, for the firm which lets or manages residential property, membership of the National Approved Letting Scheme (NALS) provides a badge of accreditation. A firm's principal, a partner or a director must belong to at least one of the following:

- the Association of Residential Letting Agents (ARLA)
- the National Association of Estate Agents (NAEA)
- the Royal Institution of Chartered Surveyors (RICS).

Planners

Professional planners belong to the Royal Town Planning Institute (RTPI). Some run or are employed by planning consultancies. Others are employed in a variety of organisations, including:

- local planning authorities, eg in development control
- government departments, eg in the planning inspectorate
- real estate consultancies
- colleges and universities
- retailing and other companies with expanding property portfolios.

Quantity surveyors

The quantity surveying professions have largely amalgamated with the Royal Institution of Chartered Surveyors. The traditional role has been in the specification and measurement of construction for pricing purposes — building economics. They have developed interests in offering services in:

- construction cost planning and management
- whole-life cycle economics of buildings
- management of capital allowances in building construction.

Valuers

Professional valuation and appraisal is mainly covered by members of the Royal Institution of

Chartered Surveyors and the Institute of Rating, Revenues and Valuation. Their work covers the spectrum of types of property — from agricultural land and buildings to zoos. The purposes for which valuations are required include:

- open market transactions
- national and local taxation
- landlord and tenant
- development decisions
- compulsory purchase and planning compensation
- insurance
- probate.

Money laundering

A property transaction is sometimes used as a vehicle for laundering the money — proceeds from criminal activities. Firms of solicitors and certain other professionals, eg estate agents, are required to have in place an anti-money laundering compliance officer who receives details of suspected or known activities from a partner, director or member of staff. The compliance officer is required to inform the authorities if there are grounds to believe the information is verifiable.

Leadership and management development

Every practice requires a leader or leaders and also managers. Specialist managers may be imported but sometimes the growing practice cannot support non-cognate specialists. It follows that the practice must seek-out recruits of the right calibre or must develop candidates from within the ranks. CPD will be required of individuals, such as:

- orientation studies or experiences for business leadership
- specialist training in the management of a particular business function.

Apart from detailed or specific management techniques much of this development work will be concerned with intra-personal skills, society and the firm's business environment.

Organisation or career development

Many organisations in the property world are growing or declining within a mass of broadly steady-state bodies. Perceiving change and learning how to deal with it must exercise many in the property industry. They will do this for their own organisation or for client entities. The strategies adopted by managements include:

- organic growth
- acquisition of going concerns followed by absorption
- mergers of two bodies
- management buyouts

- disaggregation of larger organisations.

Apart from the first, most of these strategies involve processes of bring together cultural entities and systems entities. In the case of international mergers the impact may be one of intra-cultural change as well as the more usual change of business cultures.

In all these instances the professionals involved have to quickly adopt new ways and be seen to do so. Some academics may see or have in the recent past seen opportunities to be involved in the development of a new niche profession, such as:

- facilities management
- environmental land management
- alternative dispute resolution
- conveyancing
- planning supervision
- emergency and rescue property management.

At the same time academics, perceiving the changes intuitively or by marketing research develop marketing mixes for the emerging niche professions, including those noted above.

Engaging a Professional Consultant

10

Aim

To explain how professional consultants are engaged

Objectives

- **to find a consultant**
- **to determine whether a consultant is qualified and experienced**
- **to show what insurances a professional consultant needs**
- **to review the nature of disputes**

Introduction

In all fields of the property industry professional consultants are appointed:

- to act on behalf of clients *vis-à-vis* third parties, eg to negotiate
- to proffer advice on remedial or other work which is required, eg to draw up plans, or
- to determine a dispute, eg as an arbitrator.

Appointments follow a three step need or involvement by a prospective client:

- first, the client needs to appreciate that a concern or matter warrants the involvement of professional advisors
- second, the client needs to know what one or more specialists are involved
- third, the professionals need to be identified and selected.

The general approach to selection will be based on criteria which apply in all situations, although much will be taken for granted in a selection involving a relatively small job.

Finding a consultant

Professionals are either individual sole practitioners or in a consultancy firm or company as principals, partners, directors or staff. Traditionally, the legal persona for a consultancy is the partnership with unlimited liability for the partners. They are jointly or severally liable for any liability. Some consultancies now operate under the guise of companies, being quoted in some instances.

Professional indemnity insurance is sought to ensure sufficient funds to meet any client's loss due to breach of contract or negligence; it also removes the danger to the personal assets of the partners. A staff member is not normally personally liable, ie under the employer's vicarious liability for the wrongdoing of their staff. However, where the employer ceases to practice and afterwards an employee's negligence comes to light (which would normally have been covered), the client may have grounds to claim against the employee.

Work: recognising the problem

However, the first step to finding the appropriate professional is to identify the work that needs to be done. This may seem obvious but it does not follow that a lay person will necessarily appreciate the problem. In some instances a suitable diagnostic investigation by a general practitioner in the field will be required before the need for a specialist is recognised. The client can then seek a suitable specialist or firm.

Criteria for selection

Both large and small jobs will require a consideration of the criteria listed below although for the small one-off job the amount of detail needed will be considerably less than for large construction projects.

The experienced client with a sizeable project and who knows the field is likely to have knowledge of suitable professional firms or individuals. The selection may best be based upon objective criteria which will probably depend on several matters, including:

- availability for the duration of the work
- quality of personnel to manage the work
- concurrent commitments
- resources of the firm that can be devoted to the project
- provision of suitable references
- working capital to see the job through
- price.

Finally, the approach to the selection should be considered. Possibilities include:

- personal recommendation
- recommendations from a professional body
- tendering.

Where the client wishes or needs to show objectivity in the choice of consultant, perhaps as a trustee, the last method may be suitable.

Personal recommendation

A recommendation from a person who has already employed a particular consultant for the kind of work needed is probably helpful but, of course, it is likely the general practitioner would be able to recommend someone. If possible a conversation with one or more previous clients on the quality, time and budget controls of the professional may reveal useful insights. At the same time any opportunity offered to inspect work should be appreciated as a chance to "see for oneself".

Professional bodies

A professional body may offer a service of providing the names of their members who are suitably qualified and experienced in a particular type of professional work. In some instances the professional body will have accredited a particular academic or vocational qualification. Experienced members with such a qualification gain recognition as being competent in the field.

Appointment under a contract

Contracts used in the property industry often provide for particular appointments to be made by reference to a third party. For instance, for a dispute concerning a landlord and a tenant, the lease may provide for arbitration and that in the absence of agreement on the appointment of the arbitrator, the parties should refer to the President of the Royal Institution of Chartered Surveyors or to the President of the Law Society.

Invitation to tender

An approach which is suitable for a large project or for an appointment at arm's length is the invitation to tender. Here, a formal approach which is usually conducted between groups of professionals is envisaged.

The experienced professional

It is essential that a professional adviser has the knowledge and experience to carry out the work which the employer specifies or seeks advice as to what needs to be done. The prudent employer who wishes to engage a consultant to give advice or undertake a service might consider the matters outlined in Box 10.1.

In a particular case some items will not be relevant; nevertheless it would be appropriate to assess which may be important and discuss them with the firm's representative before entering into a contract.

Box 10.1 Matters to be considered before engaging a professional consultant

Membership of professional bodies	• involves compliance with mandatory professional standards • justification of any non-observance of professional guidance • compliance with continuing professional development (CPD) • compliance with regulations covering the conduct of business, eg insurance, and care of client's money
Educational and training	• education and training to a relevant degree or professional examination standard • recorded evidence of CPD or practice certification
General and specific experience	• evidence of experience in the field, supported by qualifications • a portfolio of work done or available to be seen • references or word of mouth recommendations • view work previously successfully completed
Insurances	• consider the insurances which may be needed (see Box 10.2)

Insurances

Box 10.2 Insurances which may be needed by a professional consultant

Insurances	Professional indemnity	• seek sight of the policy and certificate of insurance • note of any limitation to the types of work covered by the policy • note any limitation on the value of the work covered
	Client's money	• professional bodies normally require cover of client's money
	Third party	• such cover may be required under statute, eg motor insurance, or contract
	Public liability insurance	• it may be desirable to have PLI
	Employer's compulsory	• employers are statutorily bound to cover employees against injury at work • note any conditions as to health and safety involved in the proposed work • consider lone working standards
	Keyman	• consider the impact on the work should a particular consultant become ill, die or depart
Personal qualities and skills		• For work involving personal interaction, the individual may require interpersonal knowledge and skills, eg leadership or project management

Disputes and differences

Professional in dispute resolution

Building and other contracts for services usually contain clauses for the method of resolving any dispute between the parties. If not, the parties may find that the Housing Grants, Construction and Regeneration Act 1996 provides for adjudication to be implied in the contract. In these situations professionals are invariably called upon to be adjudicators. It is their role to settle the difference between the parties. Generally, care should be taken in choosing the professional — in the sense that:

- the adjudicator's impartiality should not be questionable
- the adjudicator has right kind and level of professional knowledge and skills to do the job.

Disputes with a professional

Most professional work is completed to the satisfaction of the client. Unfortunately, misconduct on the part a professional is sometimes alleged by a client. Commonly on two bases:

- the professional concerned has not behaved correctly in carrying out the work but not as to be negligent, ie the matter is a complaint, or
- the professional has carried out the work in breach of contract or negligently.

Complaints

If a complaint is admitted by the professional, it may be settled by agreement. If not the complainant needs to take the matter further. Unless negligence is obvious, the complaint should be handled under a complaints procedure as a first step.

Professional firms will have an internal complaints procedure which will be taken by someone other than the party complained about. Where the firm is a sole trader, the complaints procedure will be outside the office, ie with another local firm. Again the person hearing the complaint should be impartial.

If the complaints procedure fails to satisfy the complainant, the answer may lie with one of the following:

- the firm's professional body which may have a complaints procedure
- the industry ombudsman, if any
- the small claims court or following a solicitor's letter, the county court.

Breach of contract or negligence

The subject-matter of a complaint will normally be an allegation about professional misconduct which does not amount to breach of contract or professional negligence. If there is doubt about that the matter is a professional misdemeanour, ie it does not seem to be, say, negligence, a referral to the individual's professional body by the complainant may be a good place to begin. If they think a complaint is justified, they will advise that the complaint approach the firm with the complaint and then follow any complaints procedure. If the matter is thought to be negligence the professional body will probably advise that a solicitor's advice be sought.

A professional person's conduct of business should be carried out with the following attributes:

- be careful and diligent
- at least meet the standards of peers in the profession
- at least observe any mandatory requirements of the professional bodies in the field
- follow the guidance proffered by the professional bodies or at least be able to reasonably justify any departure
- work to any conditions or limitations imposed by insurance cover
- comply with any enactments touching upon the work.

Negligence is a tort where damage is caused to a client as a result of a breach of contract where the behaviour was careless and inadvertent, but not deliberate. Where there is any doubt, a solicitor will advise on whether the matter is a question of the breach and negligence. The solicitor should be able to give advice on any remedy which may be available in law (see Chapter 34).

Fraud

Where a complainant suspects that a breach of contract did not meet the attributes given in the last section and this was done deliberately, it is possible that fraud has been committed. A solicitor's advice will be needed to ascertain the complainant's position regarding damage and the matter be brought to the attention of the police authorities.

Employing Contractors

Aim

To explain how contractors come to be employed on different kinds of work

Objectives

- **to identify the roles and activities in construction**
- **to describe the different kinds of work in construction**
- **to show how contractors are identified and engaged**
- **to describe the management and implementation of contracts**

Introduction

An understanding of the property industry requires an understanding of the terms construction (or building) industry or civil engineering industry. Generally, the terms tend to be used loosely but the context frequently determines the predominant usage. For instance, the building or construction of a bridge, railway or road is within civil engineering: whereas the erection of a cinema, theatre or housing estate would be in the construction or building industry. Of course, some construction contractors may undertake work in both industries. Finally, the repair or maintenance of buildings tends to be thought of as being done by those in the building industry.

In this volume, a general approach or perspective is to deal with those in the building and construction industry as service providers, ie they are not owners of land and buildings, unless the context requires otherwise. (The main exception is the so-called builders, eg house-builders, who are in effect dealers in property where their prime function is to build for sale.)

Listed companies in property, building and construction are categorised by the financial press under several headings, as follows:

- construction and building materials (*Financial Times*)
- real estate (*Financial Times*), covering property investors, etc and professional consultancies
- construction and property (*The Times*)

- professional and support services (*The Times*)
- support services (*Financial Times*).

In this chapter, the terms landowner, client and employer generally mean the same, ie someone who owns land and engages another to carry out some work. Such a person will be a landowner, either as a landlord developer or as a leaseholder developer. If necessary, as may happen, reference may be made to a landowner as landlord, but the client is the lessee in the same property.

However, sector, specialist or niche property players (investors, developers or operators) are not covered by the above stock exchange categories but are found in such groupings as "Leisure & Hotel", "Mining", "Natural Resources", "Retailing" and "Transport".

It may be noted that many of those in construction and building not only provide services but are also property owners as investors or operators of businesses.

Roles and activities

The many roles in construction and outlines their principal activities are given in Box 15.1. The activities are pre-contract or post-contact, although some will continue through both stages. The portrayal in this box does not indicate any of the following:

- the nature of the approach which the parties might adopt to the procurement of a particular job, eg a shop fitting job, a traditional architect design or manage project
- the relationships between the various parties as shown by contract
- the communication structure, formal or otherwise, between the parties
- the kind of contractor employed.

All of these can make construction procurement complex!

Building and construction industry

Contractors make up the building and construction industry. They range from the sole trader (offering a limited jobbing service) to a large international group of companies serving several building sectors as well as industrial sectors needing civil and structural engineering services. Box 11.1 reviews the main types of contractors and the services they offer.

Most building works require several contractors and a common approach to the procurement of a building is for the employer to engage a main contractor who may have the resources to do the entire job. More usually the main contractor will appoint others as subcontractors to do particular kinds of work or specialist work. For example, the manufacturer or supplier of specialist plant or machinery may become a subcontractor, eg for the lifts in a multi-storey building. (Chapter 19 covers several ways of procuring a building in a review of the main configurations of the relationships between the employer and the professionals, contractors, subcontractors who may be appointed.)

Box 11.1 Contractors and the services offered

International group	• a group of companies where each company offers contracting services to one or more business or industrial sectors
Building company (general)	• a building contactor able to undertake building in several sectors
Building company (specific)	• a building company offering building services in a particular sector, eg industry sector or orientation) or sector with a special orientation, eg historic buildings
Specialist company	• individual company offering specialist services, eg shop fitters using another company's products
Product company	• a building product(s) company offering installation or fitting services
Specialist jobbing builder	• sole trader or small company offering specialist services, eg decorators, carpet fitters, security installations
Jobbing builder (sole traders)	• individual offering one or more building trades, eg plumbing, decorating, electrical services

---

Types of work

Construction usually refers to work on an existing building or that involved in creating a new building. It also covers the creation of infrastructure but this is usually referred to as *civil engineering*. Both construction work and civil engineering work may be further categorised into the following types:

- new build
- improvement
- repair
- maintenance.

New build

New build is, as the name suggests, a complete construction from scratch either on a greenfield site or a brownfield site; the latter may involve demolition, decontamination and clearance.

Improvements

When a building is purchased for improvement, the developer is likely to find hidden problems once the contractors begin working on the property. An improvement to property may involve new build in the form of extensions to the existing building but also includes new plant and machinery, eg central heating, where none existed previously.

Care should be taken to identify what work is new works and what is repair or maintenance, since for a business occupier or an investor the taxation consequences are different.

Where an occupier or developer improves an existing building there is an expectation that extra works due to hidden defects may arise, such as:

- asbestos may be found, requiring special treatment (see Chapter 14)
- dry-rot in timbers
- water penetration.

This perception is likely to lead to appraisals which provide for an allowance for contingencies or an extra slice of risk and profit. Also, the contractor may be required to apportion the costs.

Repairs and maintenance

Repairs are usually treated as an expense of the business for income tax or corporation tax. However, on an acquisition the treatment in the first two years of ownership may be different. Thus, any repairs undertaken during that period may be capital expenditure. If so, they would feature in a CGT computation on a disposal at a later date (where the taxpayer is an investor or a business).

Fitting-out works

A developer will sometimes fit-out a property either prior to letting or in agreement with the lessees before they take the premises. Usually, but not invariably, the developer does shell and core works which cover the following:

- the structure of the building
- the envelope for unit accommodation
- the services core
- the plant and machinery for the building services
- basic building services.

Where the developer goes further, "Category A" fit-out might be undertaken by the developer in addition to the shell and core works. However, where prospective tenants are involved, the extent of the work may vary by agreement. In offices work covers such items as:

- suspended ceilings
- lighting
- air conditioning
- raised floors
- carpeting.

In retail premises the work may include many of these items but retailers often have specific requirements and the décor and fittings are to the individual retailer's "house-style". Nevertheless, continuity of the project management and the contractor's workforce may suggest that the retailer would use the same resources.

Finding contractors

Competitive tendering

An employer is best advised to use competitive tendering for many types of construction work. For selective competitive tendering, prospective contractors are selected from a list of approved companies who have passed through a pre-qualification procedure which attests to their suitability for the work. The pre-qualification stage covers such matters as:

- evidence that the applicant company has successfully completed similar projects
- the previous work was to the standards required
- the company has sufficient resources, eg plant, machinery and workers, to start and complete work
- the construction management team are qualified and experienced in the kind of work to be contracted
- appropriate insurances are to hand and any performance bond requirement will be obtainable
- due diligence reveals that the company is appropriately financially resourced to cover the work.

Statutory or contractual requirements

In seeking contractors for work the employer may have to comply with one or more statutory or business contract requirements. (Similarly, a contractor may find that the employer requires particular subcontractor to be employed.) Examples of these conditions are:

- for boiler work with gas, the need under enactment to employ CORGI registered contractors
- for public works, the European Union requirements for publicity (see below)
- for specialist works, the manufacturer's guarantee or warrantee may be dependent on the employment of a particular installer.

Standards and trade associations

Builders have a number of trade associations; membership of which indicates recognition of the need for standards. Two such bodies which are common for small to medium contracts are:

- the Guild of Master Craftsmen
- the Federation of Master Builders.

Almost every type of material or building component has specialist contractors associated with it. Also, most have a trade association able to offer technical advice or sources of advice, as well as information on suppliers, installers and contractors.

The British Board of Agrement certifies products, eg as meeting the requirements of the building regulations. This follows work of product evaluation, testing, production inspections and the like.

National House Building Council

The National House Building Council (NHBC) provides a new house with a guarantee against defects for a period of 10 years from the date of completion. Builders with the NHBC Buildmark scheme are likely to consistently build to acceptable standards.

Multiple quotes

Having reviewed possible contractors for a job, and assuming that each builder has pre-qualified by submitting the relevant information, the client needs to choose one of them. It is best to invite at least three to submit a quote for the work, together with other appropriate information about how it will be tackled. It does not follow that the cheapest quote will be the best choice.

International companies

Public bodies in the European Union are required to advertise some jobs in the *Official Journal of the European Union* (a regularly published source of information, etc) so as to promote the work in a pan-European way, thus complying with the competition principles of the founding Treaties.

Contracts

There are several groups of standard contacts which employers and contactors use to cement the relationship between them. The Joint Contracts Tribunal (JCT) forms of contract are listed in Chapter 19. Generally, the main contract is between the main contractor and the employer. The main contractor tends to have sub-contractors for some bundles of work. Each sub-contractor will have a contract with the main contractor: not the employer. Sometimes the employer will nominate a sub-contractor and at a later stage of the construction process life may become difficult for the parties if a nominated sub-contractor fails in the performance of the work.

Self-build and do-it-yourself

Do-it-yourself, say, a kitchen and self-build operations are usually confined to those operating on a small to medium budget, including:

- homeowners
- prospective home-owners
- house make-over investors
- perhaps, farmers and others with small-scale businesses.

Generally, an owner is either a "project manager", perhaps giving a hand or a hands on operator using others when the law requires, eg a CORGI for gas work, or when the embedded skill would be too stretched. A wise precaution as botched do-it-yourself structural work, etc is a frequent note in the surveyor's survey report on a subsequent sale.

Generally, the self-build owner who starts from scratch will normally follow the stages of the development process given in Chapter 15. However, the range of development types will be:

- buy the site and build a bricks and mortar home either from an architect's design or from approved but off-the-peg plans
- buy the site and build a pre-made kit.

Disputes

A dispute between any two or more of the parties, eg between an employer and a contractor or between a contractor and a sub-contractor is likely to be settled by a reference to:

- an arbitrator
- an adjudicator, perhaps by virtue of the Housing Grants, Construction and Regeneration Act 1996
- the Technology and Construction Court or
- some other means.

Chapter 35 deals with:

- approaches to dispute resolution
- the criteria for the selection of an approach
- the conduct of each approach
- other relevant matters.

It must be hoped that the appropriate choice of approach at the drawing up of the contract will ensure that a disputed matter is handled with authority, in a speedy manner and at reasonable cost.

Buying, Selling and Other Transactions

12

Aim

To explain the way in which property transactions are handled

Objectives

- **to identify and describe the principal roles in buying and selling property**
- **to describe the methods of transfer**
- **to outline compulsory purchase procedures**
- **to indicate how to protect retained property**
- **to outline conveyancing and land registration**
- **to review how taxation arises**

Introduction

Transactions in property mainly involve two parties, the acquirer and the disposer — two terms used to suggest a wider field of transactions than the title to this chapter suggests. Although buying and selling property is the norm for a transaction, transactions involving gifts, exchanges and leasing are a few of the possible ways in which an owner transfers an estate or interest.

Transfers such as buying and selling property involve different professionals and several building trades. This chapter seeks to delve into the practices and procedures by which property passes from one person to another.

Roles and activities

Box 12.1 sets out brief descriptions of those involved in one or more of the methods used in buying and selling land and buildings. The content is somewhat general in that different approaches will require different sets of advisors. The following illustrate the point:

Box 12.1 Roles and activities in buying and selling land and buildings

Vendor/seller	• appoint one or more estate agents and provide them with relevant information and access
	• provide access for viewing by applicants
Applicant/buyer	• register with the estate agent/auctioneer
	• view the property
	• appoint valuation surveyor or building surveyor and consider their advice
	• arrange adequate finance
	• appoint solicitor to deal with the contract etc following a successful offer
Valuation surveyor	• survey and advise on value for purchase or sale
Building surveyor	• make a structural survey and report on condition and the need for repair, improvement
Structural engineer	and regulatory compliance
	• advise on potential flooding, atmospheric pollution, noise and other local features
	• arrange for any specialist investigations or tests, eg drains, electrics, and settlement of the building
	• measure the property and prepare plans for repairs and improvements
	• advise on the need for insurance
Land surveyor	• measure the land and buildings and prepare plans, perhaps with levels and contours, particularly where redevelopment is proposed
Estate agent	• survey and measure the property with the view to preparing particulars for its sale
	• prepare particulars and promote the sale
	• receive applications and arrange for them to view
	• make recommendations on any offers
	• if necessary, liaise with the solicitor in preparing the contract and then the completion
Auctioneer (generally	• prepare the auction particulars
as the estate agent)	• conduct the auction
Bank	• advise on the buyer's position regarding funding and the types of loans available and
Building society	the terms and conditions
Other financial	• liaise with the solicitor and have the funds available at completion
institution	
Solicitor	• obtain the appropriate completed Seller's Property Information Form
	• prepare contract for the sale
	• make enquiries of the Land Registry
	• make local searches
	• liaise with the bank or building society making a loan
Land Registrar	On registered land
	• respond to the solicitor's enquiries
	• register the sale
	On unregistered land
	• investigate title with the view to registration
	• register the sale

Box 12.1 continued

Local authority	• respond to requests for local searches
	• respond to request for information about council tax or business rates
Specialists	specialists may be required to inspect and test various services in a building, including:
	• foundations for settlement
	• drains
	• electrical systems (qualified, competent electrician for works)
	• boilers and plumbing systems (GORGI for gas)
	• security systems

- an estate agency, for instance, may specialise in a particular property sector, eg woodlands and forestry, agricultural land and buildings, and industrial premises
- a particular firm may not conduct auctions, although this may be an appropriate way to deal with a property, eg on behalf of trustees
- a sale by compulsory purchase will normally involve a valuer or estates surveyor who is a specialist in the subject.

Where appropriate, some participants may, therefore, undertake activities in more than one role.

Methods of transfer

Open market sale

Property is usually sold on the open market by the owner engaging an estate agent to act as intermediary. The method recommended by the agent or specified by the owner will depend on circumstances, usually involving one of the following:

- private treaty negotiations
- auction
- tender.

Private treaty negotiations

Negotiations are probably the most common means of disposal. An estate agent will normally be engaged and carry out the activities listed in Box 12.1. In preparing the property particulars the agent should take care in measuring the land and buildings so as not to breach the Property Misdescriptions Act 1991.

Auction

An auction is often the recommended manner of sale when:

- the seller wants to or is required to show that a fully open sale process was undertaken, eg the seller is a personal representative, a trustee or a public body
- the property is specialised, unique or unusual
- the extent of the market for it is uncertain
- there might be a range of possible buyers, perhaps with diverse objectives in mind.

Tender

Tenders are often invited when the seller or prospective landlord wishes to "control" the outcomes after the disposal. For instance, the seller of land for development who retains adjoining land will want to protect the retained property. Such a sale or lease might proceed by the steps shown in Box 12.2.

Box 12.2 Processing a schematic tender for the development of land

Step 1	prepare: • brief details of the development proposal • pre-qualification questionnaire • the planning brief (unless information is in the development brief) • development brief
Step 2	promote: • the proposal's brief details • invitations to prospective developers to express interest in doing the development
Step 3	receive expressions of interest from developers
Step 4	issue to interested developers: • the pre-qualification questionnaire • the development brief • an invitation to meet to discuss their considered response
Step 5	after meetings, invite one or more developers to submit more detailed bid
Step 6	select detailed proposed development and prepare: • contract for sale with supplementary documents, or • building licence and agreement for lease, or • building lease
Step 7	monitor development from beginning to completion
Step 8	finalise the documentation for the sale or lease upon satisfactory completion of the development

---

Forced sale

A freeholder may be required to sell property off the open market but where there is an assumption of a sale price at the open market value (perhaps with other assumptions). This arises under:

- the Leasehold Reform Act 1967 (as amended) — to the qualified lessee of a dwelling-house held on a long lease
- the Leasehold Reform, Housing and Urban Development Act 1993 — to the qualified lessees of a block of flats
- the right to buy — qualified tenants of social landlords enjoy the right to buy their dwelling
- the Landlord and Tenant Act 1987 (as amended) — a right of first refusal, where tenants, whose landlord is selling a block of flats, may buy it
- where a previous owner or tenant have acquired the right of first refusal, the property must be offered before it is put on the market
- an option to buy the property at the open market value
- a conditional contract, eg a sale subject to the purchaser obtaining planning permission within a limited period, say 10 years (the price being a proportion of the open market value when the condition is satisfied)
- a personal representative may be required by the deceased under the terms of the will to sell property.

Compulsory purchase

Where the property is being purchased as a whole or in part under a compulsory purchase order, the seller (claimant) will normally engage a professional valuer to negotiate, not only the sale of the property but also any entitlement to disturbance or other items of the claim. The valuer would normally be involved in any appeal to the Lands Tribunal in the event of a dispute (see below).

Buying a company

The purchase of all the shares in a company will result in any freeholds or leaseholds in land and buildings that the company owns passing, in effect, to the purchaser. The purchaser's professional advisors will normally undertake a due diligence examination of the company and advise on any problems.

Other transfers

Other occasions result in property being transferred, including:

- when an individual dies the property may be transferred by the personal representative to an heir or under the rules of intestacy
- when an individual dies without a will and there are no relatives, the property passes to the Crown under the rule of escheat
- when an organisation is put into administration, property will pass to creditors or be sold to pay debts
- in times of war or in the aftermath of an extremely severe terrorist attack, property may be requisitioned
- property may pass by deed of gift either by outright gift or by trust.

Compulsory purchase

Government departments, local authorities and other public bodies have long been able to buy land compulsorily and by agreement. As a result of the privatisation of many public bodies compulsory purchase has become more common in the private sector. Also, oil and other companies in the private sector may seek powers to obtain land and rights (easements and wayleaves) compulsorily for pipelines under the Pipelines Act 1962.

Powers

Powers for the compulsory purchase land and buildings are obtained under the Acquisition of Land Act 1980 as amended, particularly by the Planning and Compulsory Purchase Act 2004.

Compensation

Compensation is assessed under the Land Compensation Act 1961, as amended. When a property is affected two kinds of case are possible, namely:

1. where the whole property is acquired
2. where part of the property is acquired.

As a result of the Compulsory Purchase Act 1965, an owner may be able to argue that the whole property be bought.

The owner is entitled to claim compensation. The measure of compensation is the open market value of the property, together with a claim for disturbance, if any.

Disturbance may be justifiable where a property is acquired without any actual or hope of (or assumption of) planning permission, ie its open market value does not include any development value.

Where there is no development value the claim will be as follows

Open market value (as existing) + Disturbance + Professional fees

Where there is development value, the claim will be as follows:

Open market value (with planning permission) + Professional fees

In accordance with the 1961 Act, the planning permission may be an actual permission or an assumed permission (which is more likely).

Protection of retained property

Protection of any part of a property which is retained will be needed where the property owner is disposing of a part, either by a sale or a lease. Depending on the circumstances several ways are available to the owner, including:

- disposal by Building Licence and Agreement for a Lease (see Chapter 3)
- in a lease, user restrictions and other terms and conditions

- disposal subject to restrictive covenants and other servitudes (see Chapter 4)
- on compulsory purchase of part of the property or where no land is taken — accommodation works may be agreed to safeguard the property, eg a boundary wall.

Buying a house or flat

Information requirement

At present prospective purchasers of dwellings need several kinds of information, including:

- for freeholds, Property Seller's Information Form
- for leaseholds, Seller's Leasehold Information Form
- for registered property, the Land Registry
- for unregistered property, the Register of Deeds
- planning register
- property certificates
- statutory and land charges register search.

The solicitors or conveyancers acting for the parties would normally gather this information after a transaction had been agreed subject to contract.

The Housing Act 2004 provides that the seller of a home will be required to have available to buyers a home information pack (HIP) containing similar information and additional data, including a home condition report (HCR). The latter information will be prepared by a qualified and insured home inspector — one of 7500 to be recruited for the work.

Structural defects

The purchase of a dwelling is probably the most important transaction that most people will make. It is important, therefore, to ensure that everything is done to ensure that problems are minimised or eliminated before completion. The survey should reveal structural defects and other problems associated with the building.

Box 12.3 gives common defects and other matters which may affect:

- the decision to go ahead with the purchase
- what may be required of the vendor prior to completion
- what the buyer may expect to do from completion.

For a house being built, snagging should pick up many faults prior to completion of the house and any survey.

Of course, many of the points in Box 12.3 apply to commercial and other non-domestic property.

Box 12.3 Common structural defects and other problems in dwellings

Cavity walls	• check for failure of wall ties
Contaminative uses	• where remediation certificate has been issued, ensure works are complete and acceptable • check monitoring safeguards of any remediation works are in place and working
Extensions	• check extensions are properly tied in to the main structure
Fire damaged property (before completion)	• if vendor insures, ensure that insurance monies are applied appropriately or agree a settlement to cover damage
Hedges	• check high hedges not likely to result in action by neighbour(s) or the local authority
Housing orders	• check that closing or demolition order is dealt with
Listed building etc	• if appropriate, ensure listed building consent or conservation area consent is obtained or obtainable
Movement of the superstructure	• check for doors and windows which are out of alignment due to movement in the superstructure • check for broken pipes, cables and conduits
Radon	• check on the need for works (see Chapter 14) • if works done, check their efficacy
Replacement windows	• check that windows are in keeping with the style of the house • check fixings and lintel support above • check for broken seals on double glazing
Timbers	• check for dry rot and other rotting of timber • check roof timbers at eaves
Trees	• if tree preservation order protects trees ensure compliance with procedures
Settlement	• check for settlement of foundations, bay windows and new extensions
Subsidence	• check for mine shafts, dene holes, old wells, cess pits and other cavities • if the property is near the seashore check the policy for sea defence management in the area
Structural alterations	• check do-it-yourself works, particularly structural load bearing work
Vermin	• check for attacks and adequacy of protective works against mice, rats, squirrels and wood boring insects • check for wasps and hornets nests if works are to be done
Guarantees, warranties and insurances	• check wording and adequacy of any guarantees, warranties and insurances for works done by contractors • check the expiry date of the guarantees etc
New building	• ensure snagging is undertaken by a competent person • check the work has a valid certification by the National House Builders' Council, ie Buildmark • check any guarantees, warranties and the like

Conveyancing

The formal creation, transfer of extinguishment of an estate or interest in land (and buildings), ie conveyancing is usually the responsibility of a solicitor or licensed conveyancer. The work involved is shown in the schematic Box 12.4.

Box 12.4 Schematic representation of conveyancing in general

Step 1	seller's conveyancer	• obtains office copy entry from Land Registry or other proof or title • prepares draft contracts
Step 2	buyer's conveyancer	• makes searches of land charges register and other official data (unless supplied by seller's conveyancer) • receives draft contract and vets it for approval • enquires of the seller about the property • completes enquiries about the area (unless done by the buyer)
Step 3	both conveyancers	• obtained signed contracts from their clients • exchange contracts • deposit paid by the buyer • ensure continuity of insurance cover for the property (note of buyer's interest on the policy)
Step 4	buyer's conveyancer	• seeks confirmation of proof of title by enquiries at Land Registry or title deeds (is unregistered) • clarifies any uncertainties • rights of others searched at Land Charges Registry (unregistered property) or Land Registry (registered property)
Step 5	both conveyancers	• agree final draft • obtain approval and signatures of clients • registers the unregistered property with the Land Registry • complete the purchase • buyer pays the money due and receives the title documents
Step 6	buyer's conveyancer	• registers the title at the Land Registry (if seller's title registered) • formally notifies the Inland Revenue
Step 7	Inland Revenue	• assess for any stamp duty land tax payable by the buyer • record the transaction for statistics
Step 8	buyer (through conveyancer)	• makes payment of any stamp duty land tax and land registration fees • apportionment, if instructed, of business tax, council tax and metered charges
Step 9	seller and buyer	• makes business records for annual financial accounting • makes records for capital gains tax (or income taxation — dealer) • considers availability of any capital allowances (buyer from seller) balancing charge or allowance (seller) • cancel insurance cover (seller) ensure continuity of cover (buyer) • seller may find CGT payable (if an owner occupier of house used in part for business, part let or with large garden

Land registration

Land Registry

The Land Registry is a government agency empowered under the Land Registration Acts, particularly the 2002 Act which has repealed much of the earlier legislation, eg the Acts of 1862, 1925 and 1936.

Registered information is guaranteed as accurate and should a person suffer loss as a result of an error, a claim for compensation may be made.

Proof of title

When a transaction in property has been agreed, the person acquiring it will want confirmation of ownership. Proof of the seller's ownership of a property is demonstrated in one of three ways, namely:

1. the estate held in the property has been registered by the Land Registry (it will have a reference number or title)
2. where the property is not registered, title to it can be established from documentary evidence, the "title deeds"
3. where possession of land is apparent or claimed, but the ownership is not registered and there is no documentary evidence, title may be confirmed by adverse possession being established.

Registered land

Registration of land has been both compulsory and voluntary for many years. In most instances the property will have been registered by the Land Registry at some earlier time either compulsorily, eg at the time of a previous transaction, or voluntarily at the behest of the owner.

A reference number or "title" is given to each freehold and leasehold in the land.

The local District Land Registry office will have details which are open for public inspection for a small fee. The details which may be found are:

- the proprietor (owner) of the land
- mortgages affecting the property
- covenants on the land
- easements burdening the land (but not all types)
- leases over 21 years
- for transactions from 1 May 2000, the price paid for the property.

After the sale the buyer's solicitor will register the new proprietorship at the District Land Registry. Fees will be payable for the service.

Unregistered land

Where a property has not been previously registered, the solicitor for the person disposing of it will have to be instructed to prepare a proof of title prior to contract. An investigation of the title deeds will be made and the appropriate abstract prepared in advance of registration by the Land Registry.

The Land Registration Act 2002 encourages owners of unregistered land to voluntarily register.

Electronic conveyancing

The 2002 Act provides for a speeding up of conveyancing by empowering the Land Registry to develop on-line conveyancing.

Taxation

This section is a reminder that taxes may arise when property is sold or otherwise transferred. Part 8 covers much of the detail in the situations alluded to below.

Lifetime disposals for money or other consideration may result in the owner being liable for capital gains tax and the buyer being subject to stamp duty land tax. There are, of course, many exemptions and reliefs. For instance, the owner-occupier of a dwelling is not usually liable for capital gains tax.

Inheritance tax may arise on the death of an individual, even on property given away in the seven years prior to death. So-called favoured assets in the estate are either exempt or take some form of relief provided they pass to beneficiaries. A sale of a favoured asset after death may result in a charge to inheritance tax.

Also, when property is leased and the landlord receives a premium a liability to income tax or corporation tax as well as capital gains tax is likely unless an exemption or a relief is available. A business tenant who pays a premium which is taxed in the hands of the landlord may be able to claim relief against business profits by way of a notional rent.

It follows that the proposed disposal should be carefully appraised for any taxation consequences which may result. Where an individual or organisation has a business plan for a property, it would be prudent to cover the taxation consequences on a disposal of any earlier works to the property — capital allowances may be available to the purchaser and any capital gains tax may be less as a result of the cost of the works being "enhancement expenditure".

Part 4

Development of Settlements

Settlement and Community Development

13

Aim
To establish the ethos of sustainability and community development

Objectives
- to explain how infrastructure is provided in settlements
- to describe the approaches and organisations involved in the development of settlements
- to describe approaches to town and community management

Introduction

A broad perspective of development process requires consideration of the always evolving aggregation of future developments in society. In other words, while each individual development is an example of a development process, society requires action to plan for and provide adequate infrastructure, ie facilities for the likes of:

- education — new or extended schools, colleges and universities
- flooding — improved planning and new infrastructure (see below)
- health and care — new specialised clinics, practice centres and hospitals
- housing — housing to "fit" projections of the rates household formation (see below)
- information and communications technologies — creation of cyber-accommodation, cabling systems and networks of towers for telecommunications
- transportation — creation of facilities for regional and national airport developments, new fast railway systems, container ports and motorways
- leisure and recreation — expansion of facilities for public recreational access and participation in sport and cultural activities.

Considerations such as these take place at regional or local levels within an evolving national planning system. This may be seen as the development processes at the two planning interfaces, namely:

- national and regional bodies
- national (or regional) and local bodies (see Box 13.1).

The relative importance of the regional bodies to government in planning and development is yet to be established; the lack of support for the regional assemblies suggests that these bodies will not come into being in the short term, if ever. At present in England the regional structure is represented by elected lower local government tiers and others. Nevertheless, the government is giving powers to the regions, eg housing strategy.

Box 13.1 Distribution of town and country planning powers

National government
- ODPM
  - planning inspectorate — determines planning appeals
  - ODPM deals with major projects, as reserved call-in applications
  - promotes guidance on regional planning, minerals planning and general planning
- DTI
  - develops wind farm development policy
  - regulates oil and gas exploration and production
- DCMS
  - listing of historic etc buildings
- DEFRA
  (see Chapters 31–33)
  - rural affairs
  - environmental matters

Regional development agencies
  - develop regional economic and development policies
  - feed into regional planning guidance

London Assembly
  - possibly, regional housing strategies

County councils
  - prepares mineral plans

Unitary authorities
  - forward plans
  - development control
  - enforcement
  - special controls

District councils
  - forward plans
  - development control
  - enforcement
  - special controls

Town and parish councils
  - planning observations on plans, planning applications
  - plans and manages local amenity initiatives and programmes

---

The next chapter concentrates on the criteria for sustainable and acceptable development of a single plot or area.

Contexts of national and regional policies

National planning concerns turns on the place of the UK in the European and world economies and on the state of the country, ie the latter in terms of economic, social and political factors. A comprehensive study of the strengths–weaknesses–opportunities–threats (SWOT) of national planning may hinge on such matters as those shown in Box 13.2. The items under the headings in the box represent part of an attempted macro-view of the underlying changes taking place in society. All have some impact on the property industry.

Box 13.2 SWOT analysis on example factors in national economic planning and hence town and country planning

Strengths
- capability in innovation, eg information and telecommunications technology, biotechnology
- capacity and growth in technological development
- improving infrastructure
- financial industry and service industries generally
- service industries are strong, particularly finance and insurance
- population growth
- time-zone location of the London financial markets

Weaknesses
- in technological transfer to manufacturing
- investment short-term perspectives
- problems with work-lifestyle balance
- management and worker training and development
- poor productivity against peer nations
- high rate of household formation
- health and safety in construction
- shortages in educated and trained workforce
- foreign languages skills and abilities
- manufacturing industry losses to foreign competition

Opportunities
- potential of alternative energies
- technological growth has potential in many fields
- substantial population growth of EU consumer population — giving potential for exports
- changes to the education system
- heritage and tourism from overseas
- Olympic Games — to London 2012

Threats
- global warming
- international terrorism and organised crime
- depletion of the North Sea oil and gas field
- growing strength of manufacturing output, technological development and service industries in India, China and elsewhere
- work-lifestyle balance
- attitudes generating investment short-termism
- growing consumption of world's natural resources
- aging population

Hierarchy of settlements

Existing settlements range from hamlets to conurbations; they are in a flux of growth, steady-state, stagnation or decline. Successive governments have initiated, developed or built upon programmes and projects to arrest stagnation or decline. In Chapter 31, Box 31.1 gives examples of major controls, programmes and projects from the middle of the 20th century which have impacted on the property industry. Of course, almost all settlements are situated in the countryside and planning seeks to address rural issues and concerns.

Old and new settlements

Existing settlements range from huge conurbations to small hamlets and isolated farmsteads. An expected expansion of the population or the demand for homes of an existing settlement requires growth of the settlement. This follows a relatively predictable process, namely:

- conversion of existing buildings
- infilling existing vacant sites
- redevelopment of existing buildings or derelict sites
- clearance, demolition and redevelopment of derelict areas
- development of greenfield sites on the periphery.

When a new settlement is proposed its scale ranges from a few buildings to a new village, town or city. In all of these development contexts the "expanders" will meet thresholds which must be overcome if the development is to proceed and be successful.

For instance, it was decided in the 1950s to create several new towns. The threshold was the organisational structure needed. The government designated about 30 development corporations, each with a remit to create a new town. Briefly, each development corporation prepared a master plan. They duly bought land and buildings, constructed infrastructure and endeavoured to create a balanced community with houses, shops, factories and leisure properties. Each new town had a hierarchy of road networks and cycleways linking residential areas to areas for industry and commerce as well as the town centre. Shopping was provided in a hierarchy of local, district and town centres, generally before the advent of the out-of-town shopping centre. Milton Keynes is a prime example of a new town.

Small towns and villages

Several small towns and villages have been created since the 1960s, when New Ash Green in Kent was created. In more recent times Poundbury was created by the Duchy of Cornwall. Each reflected the prevailing ethos of modern development of its time. More recently, it has been announced that a development framework is being formulated for a new settlement at Telford. In these instances, land ownership was or may be an issue. In the first instance, the private developers had to acquire land. In another, the land was not a threshold since it was already owned.

Single buildings

New isolated individually owned buildings in the countryside are relatively rare but are sometimes built for agricultural purposes, country houses, military purposes or telecommunications. The cost of services is high unless they are independent of the national or local networks.

Infrastructure and thresholds

New settlements require the planning, design and construction of infrastructure to support the occupied (and to be occupied) buildings. It usually takes several years, if not tens of years, to acquire the land and construct a complete new town. Each new development must have essential infrastructure and utilities, together with readily accessible services, eg shopping, before it can be occupied comfortably.

Thresholds options and their costs

The planning of a new settlement or the expansion of an existing settlement throws up thresholds. These are both natural and unnatural barriers, eg rivers or railways respectively. The thresholds must be stepped over or by-passed to enable the development to proceed. They may be classified as shown in Box 13.3.

Box 13.3 Thresholds in the development of settlements

Organisational	• happens when an existing organisation is not able to undertake the development, eg the new town development corporations replaced local authorities
Tenurial	• compulsory land acquisition powers are needed when land is held spatially in many ownerships and the owners are unwilling to sell • restrictive covenants and other servitudes are sometimes a threshold
Heritage	• an ancient monument or an environmentally sensitive area is a barrier which on cultural grounds must be weighed against the benefits of development
Area	• a river's mudflats must be drained and walled • a local road network must be constructed before an area is developed, eg a new road layout of x kilometres for every 10 hectares of housing
Physical	• a cliff may need to be surmounted or a hill removed • a seawall to protect a town expansion
Linear	• a road or railway, will cost so much a kilometre
Population	• a lack of population may hold up the construction of a shopping centre
Stepped	• a new school for every 5000 increase in the number of housed population

---

It is likely that each threshold can be overcome by one or more infrastructure solutions or options. For instance, the crossing of a river to enable development on the other side may be achieved by adopting at least one of the following:

• a tunnel and mode of transport facilities

- a bridge and mode of transport facilities
- a ferry facility and mode of transport facilities.

A possible mode of transport for a ferry facility could be:

- helicopters or aircraft
- boats (carrying the next two possibilities)
- vehicles or on-foot travellers
- passenger trains or buses.

Modes for the bridge or tunnel could be vehicles, on-foot, trains, buses or some combination. Finally, the ferry could be by submarine! Also, the body needed to develop the crossing could be a private sector company, local authority, a trust or a utility corporation.

Of course, the volume of traffic (of all kinds) and the various capacities may limit the choices but an important factor will be the cost. Here, it is not only the capital cost which should be considered but also the running costs, ie a life-cycle cost of each option.

Highways

The Highways Agency has executive powers over the national highways system. It is responsible for:

- strategic planning of the motorways and other road systems
- creation, repair, maintenance and policing of the motorways
- the development of policing, health and safety policies for the system
- overseeing local authorities acting as agents for the repair and maintenance of other highways.

Local highway authorities have powers and central government funds for highway maintenance. Traffic management was added to the remit by the Traffic Calming Act 1992 (as an amendment to the Highways Act 1980). However, it seems that the funds are not ring-fenced so maintenance tends to be neglected. This has happened to such an extent that claims from motorists and others for loss or damage has increased considerably and now takes a fair proportion of the budget.

Sustainability issues

Coastal protection

The Wash and other areas of low-lying marshes near the sea or inland have been drained by our forefathers who built protective barriers for resulting reclaimed agricultural land, villages and towns. Sea walls and other barriers create threshold problems and generate sustainability issues in the face of rising sea levels. Thus the Environment Agency and other bodies have evolved sustainable policies to deal with the high cost problems of coastal protection. For instance, the Environment Agency and others have evolved the following policies in managing the national budget for sea defence, namely:

- managed realignment
- managed retreat.

There is a programme of works to protect valuable land and property with improved existing and new defences. Also working with English Nature and others, the policy of managed realignment allows mudflats or salt marshes to develop on poor quality land in front of new defences protecting higher ground which is more inland. With a constrained budget, in some areas a policy of managed retreat is adopted. In such areas the sea is left to find its own equilibrium and the loss of land and buildings in these areas is inevitable. Of course with the prospect of such loss, the private sector insurance industry does not issue cover and as yet there is no policy for a compensating public sector fund — those who lose home and livelihood are normally eligible for rehousing by the local housing authority.

Energy

Nationally, energy is supplied to energy users in property from the traditional sources of coal and its products, electricity, gas, hydro-power, radioactive substances, oil and combustibles, eg timber. Generally, sustainability is not highly developed in the traditional energy industry since almost every traditional supplier lacks opportunity. Thus, apart from hydro-power and timber the traditional sources are not "renewable". It was the 1973 oil crisis which started the quest for new renewable energy sources and initiatives on energy management. Efforts have been increased since the Rio and Kyoto Agreements on sustainability.

As a result a number of renewable energy sources are being developed with the support of government taxation measures and financial incentives. The government's policy is to build the renewable energy sector from about 3 or 4% now to 10% by 2010 and 20% by 2020. It will, of course, be considered sustainable thereafter. Box 13.2 shows a variety of traditional and renewable energies. It shows the possible impacts they may have on the property industry.

Various taxes and taxation reliefs and financial instruments have been used to encourage the management of energy, including:

- aggregates levy — leading to more recycling of construction waste
- the climate change levy — inducing more energy efficient generation
- congestion taxes — increasing reliance on public transport and cycling (particularly in London)
- landfill tax — leading to more recycling and improved design to avoid waste.

Environmental impact of development

Sustainability concerns arise when developments are proposed and a planning application is prepared. Local residents and others usually object to a proposed development which is expected to cause harm to the locality, particularly if some heritage feature — natural or built — is affected. Of course, every development will have some impact on the local environment. In most cases the simpler special controls which the local planning authority has will be sufficient to evaluate the development and address concerns; but in the case of large development proposals a procedure, known as the environmental impact assessment (EIA), will be invoked. The Town and Country Planning (Assessment of Environmental Effects) Regulations 1988 (SI 1988 No. 1199) (as amended) provide in schedules 1 and 2 developments for EIAs. For schedule 1 developments the EIA is mandatory but schedule 2 developments are covered at the request of the local planning authority.

Box 13.4 Possible impacts on the property industry of energy sources

Coal mining and processing	• has declined markedly in last 30 years • left a legacy of derelict areas • some subsidence problems to be expected • hitherto a high carbon output (but technology has evolved to clean gaseous fumes in the flues of coal burning power stations)
Electricity generation (see below for electricity transmission)	• nuclear generation in decline (see below) • co-generation schemes with coal and other combustibles being developed
Electricity transmission	• increased provision of grid from remoter areas required for transmission of renewable wind-based electricity • possible fears for children's health
Fuel cells	• devices for conversion into energy • allow buildings to be non-reliant or independent of other energy systems
Heat exchangers	• enabling the recovery of waste heat from other energy sources
Gas and natural gas	• traditional manufacture declined substantially • natural gas from North Sea in decline • terminals need to be built for imports of natural gas from Algeria, Russia and elsewhere
Nuclear power generation	• being phased out (at present) • Magnox power stations are to be decommissioned • from 2005 the management of the nuclear legacy, eg cleaning up sites, is to be addressed by the Nuclear Decommissioning Authority (see the Energy Act 2004)
Oil	• long term decline of the North Sea oil and gas field • redundant rigs and other facilities will need to be recycled • re-employment training and job opportunities will needed in settlements dependent on the oil economy • land-based oil industry quietly developing in the South-East Region
Solar	• a system of energy generation which converts the sun's energy to electricity • heats water for use in buildings • variable output, may therefore need a back up system • reflective pipes may be used to transfer sunlight into attics, cellars and other dark accommodation
Timber and other biomass energies	• requires power generators using coppice or farm grown crops which may be burnt for energy conversion • oil or gas may be obtained from some crops for heating or vehicular movement
Wave and tidal power sources	• utilises the energy from waves or changing levels of the sea • possible renewable energy source • prototype generators being developed or coming on stream
Wind farms	• major on shore and off shore developments are in hand in phases • transmission by the national grid from remote locations may be costly and unsightly • strong opposition from local and national groups to local projects

Flood control

With the prospect of global warming continuing, it seems that violent storms and heavier bouts of rainfall will become more severe. For some settlements, flooding due to rain water rapidly draining off catchments is likely to be persistent. Local planning authorities, the Environment Agency and local internal drainage boards are the principal bodies which control development and ameliorate the effects of flooding. Measures taken include:

- forward planning — so that flood plains are not developed with housing and other vulnerable land uses, eg industry
- building design — to reduce the impact on individual properties of flooding, eg raised floors, porous hard surfaces to drain off rainwater and drainage ditches and ponds
- flood defence infrastructure — continued development of waterways to take away water, water storage lakes and ponds
- more publicity for the recognised warning system.

The Environment Agency has produced maps which enable a property owner or occupier to pinpoint the degree of risk of flooding and the probability of an event occurring.

The insurers for the property industry have become more concerned, wanting to be assured that the authorities are planning development away from vulnerable areas and are carrying out works to improve vulnerable areas. It seems that if progress is not forthcoming, the owners of properties who are at risk will find it more difficult, if not impossible, to obtain cover.

Town management

A feature of development policy has been induced by the growth in motorised transport, namely out-of-town shopping centres, designer outlet centres, retail parks and business parks as well as major leisure centres. The result has been the decline of the traditional town centre as the main shopping place. Where town centres are unique they survive in another guise, as a tourist centre or a market town centre. Addressing this issue has resulted, conceivably, in the development of town centre management. In fact the expansion has been accompanied by the growth of the Association for Town Centre Management (ATCM).

Town management has developed in the last 20 years or so. Here it is taken to be an emphasis on town centre management rather than the town has a whole, nor is it taken to include "community development" (see below).

It requires an individual or group who can develop a mission for the town centre and a business plan. The emphasis for the manager is to pull together the resources with which the various parties in the town can achieve the mission.

In forming the mission and the programmes to achieve it, the town manager will need to address a variety of issues. The emphasis and allocation of resources will depend on the town's key strengths and the problems which the town is experiencing and the priorities which the stakeholders have for meeting them. The programmes are likely to focus on such matters as:

- traffic management and parking policies
- retailing
- employment opportunities

- after-hours entertainment and facilities
- tourist attractions and facilities
- leisure and entertainment for the resident population
- possibly, a business improvement district.

Business improvement districts (BIDs)

The business improvement district is intended to provide projects to enhance the area of the BID. Some 22 BIDs were set up as pilot schemes by the Association of Town Centre Management, local businesses and other bodies. The projects are likely to cover matters indicated in the last section.

The Local Government Act 2003 provides the legal framework for BIDs, including a BID levy (business ratepayers paying) and the billing authority having a veto. A proposal is voted on by the persons who will be liable for the BID levy. If passed the aggregate amount goes into a BID levy fund together with any other contributions. The fund is then spent on the proposed projects.

Community management and development

Development of a community is not only a question of appropriate infrastructure and buildings but touches commitment, relationships and spirit — wanting to be involved and having the opportunities. Building a good and strong community for involvement will include the built environment, therefore, this section looks at many attributes of a thriving community and many of the points will feature in the roles and activities of those in community management and development. A community's local authority is required by the Local Government Act 2003 to prepare a community strategy setting out a mission for the community, relating to partners and stakeholders. Also, under the Planning and Compulsory Planning Act 2004 there is provision for the local planning authority to show how it has taken community strategies into account. Spatial development is a special remit for the local planning authorities to take into account.

The 2004 Act has emphasised the role of the individual in the community in the processes of the new planning regime. The Act requires the local planning authority to prepare a statement of community involvement (SCI). The SCI is not community involvement itself but may induce awareness of the need to generate polices for more involvement.

Amenity

Amenity is the attractiveness to users of the area in terms of:

- open space (parks, sitting areas, shade and shelter under trees and hard coverings and street furniture)
- management of road sweeping, street lighting, litter, graffiti and obscured road signs
- ground maintenance of amenity areas, road kerbs, bollard areas and planted areas.

Trouble shooting in community development is fraught with such problems as:

- the various relatively minor matters which fall between different administrations

- the items left out of specifications and so fall between the host of different contractors employed by the different administrations
- finding out who does what
- the time taken to see results.

Anti-social behaviour

The government's thrust against anti-social behaviour is gaining in momentum under the joined-up government work and legislative measures against community crime, drug abuse, graffiti drawings, high hedges and vandalism.

Emergency services

The Fire and Rescue Services Act 2004 draws together various services concerned with emergencies into fire and rescue authorities. In due course, a Fire and Rescue National Framework will be available. The functions covered by the fire and rescue services, are:

- fire safety
- fire-fighting
- road traffic accidents
- emergencies.

Green transportation systems

Sustainable principles suggest that settlements should be largely self-contained or self-sufficient. This may be an impossible ideal perhaps but one which should be evaluated and practised, if possible. As far as green transportation is concerned the following may need to be considered:

- vehicle, cycle and pedestrian separations (to increase health and safety and to reduce burden on health and mortality services)
- contractors collecting categorised recycled materials at source (so as to reduce transportation energy)
- encourage sharing schemes for motor vehicles (to reduce movements)
- cycleways and cyclists' changing and repair facilities (to reduce vehicular usage and to reduce burden on the health services)
- congestion taxation (to reduce congestion, accidents, to increase cycling, etc and to raise funds for public transportation services).

Information and communications technology

The changes in information and communications technology (ICT) are changing the scope and use of buildings, including:

- specification of ICT facilities in new proposals is taken for granted in all sectors

- specification of residential accommodation is becoming increasingly "smart", (to include controlled access, audio-visual entertainment system, heating, lighting, security and inclusive vacuum cleaning)
- in commercial buildings, hot-desking is increasingly common
- home-working is increasing and is changing the desired specification of the ideal home
- mobile telephones have reduced the need for public fixed landline facilities.

Waste management

The Environmental Protection Act 1990, as amended, provides for waste management. Household waste must now be collected in such a way that it can be separated for recycling. Also, recycling centres now provide for household waste to be categorised into containers for, say, aluminium, asbestos, books, cans, cardboard, clothing, foil, furniture, glass bottles, green garden waste, hardcore, junk mail and paper, metals, newspapers and magazines, oil, paint, shoes and untreated wood. None of this material need be consigned to a landfill site. Although the rates of recycling are rising, they do so slowly.

Of course, the first step in waste management is to reduce waste, eg by thinking at the design stage to reduce packaging. The second stage is the immediate re-use of complete items or parts of items without the need for waste to enter the recycling centre, eg encouragement of charity shops, centres for the deposit of books for free exchange, boot sales and recycling operations of ICT equipment, for instance. As a result of EU directives, some manufacturers must now comply with waste recycling for specified products.

Some in the property industry have developed policies on waste management which saves, for instance, on raw materials, transportation costs, and the need for landfill space. These include:

- the demolition of buildings and structures with the recycling of waste as a policy
- the use of recycled materials in new buildings, eg architectural salvage centres
- programmes for waste to be used locally to produce alterative products.

Construction is becoming "greened" as some designers and contractors adopt policies and practices which are sustainable. These may include the specifications for the following:

- designing for energy-saving, eg solar lighting and heating
- designing for water management
- obtaining new materials or components which are provided from sustainable sources
- formulating a site waste management plan
- using recycled components or materials, perhaps after they have been processed.

The next chapter gives a profile for a sustainable building (see Box 14.1).

Where appropriate, a local development document will be prepared on waste recycling by the local planning authority. (Local development documents are provided for under the Planning and Compulsory Purchase Act 2004.)

Regeneration

Much of what has been considered above is within the topic of regeneration. This section is intended to draw attention to the policies, programmes and projects which have been proposed and are being implemented under a variety of partnerships comprising inter-departmental offices, agencies, departments of local authorities, housing groups and other stakeholders. Huge amounts of public funds are promised for the regeneration of many towns and cities as well as areas of conurbations. The underlying drivers for the regeneration are varied but include:

- the need for economic growth in the North of England
- the need to meet the demands for the accommodation of economic growth in the South East Region of England
- the treatment of areas of no demand for local housing.

The remainder of this section indicates four of the programmes involving regeneration. There are many more! Nevertheless, they cover a vast area and give a flavour of the scope in the field. The four are:

- the Northern Way Growth Strategy — covering 10 work streams and eight city regional development programmes, it seeks to deliver a substantial growth package to three regions
- the Urban Development Corporations — several UDCs have been set up to deal with areas requiring regeneration, eg Thames Gateway
- the Urban Regeneration Companies — numerous local companies have been created in cities and towns to effect urban regeneration with funds from public and other sources
- Housing Pathfinder Projects — English Partnerships and others have addressed areas of low demand for housing or areas of abandoned housing with a number of regeneration projects.

Rural development

The principal concerns of rural communities hinge on the following:

- lack of opportunities for work in many countryside areas
- the take up of second homes by towns' people
- withdrawal of many services, such as shops, post offices and public houses
- reduction in the rural transport network and less frequent services
- lack of affordable housing for local people
- the decline of the small farm as a viable business.

Some effort is being made to address these matters. Rural planning policy has always had the agricultural condition on planning permissions for smallholdings and the like, eg the use must be agriculture and the occupier must be local. More recently, the policy of more extensive diversity has been permitted for redundant agricultural buildings. A planning policy statement is available on sustainability in rural areas.

Finally, financial business rating relief is given to rural shops and former agricultural buildings.

Performance of public bodies

Almost all aspects of the above public sector functions are subject to a comprehensive performance assessment by the Audit Commission, together with other bodies concerned with best value and performance targets. An overview of comprehensive performance assessment is given in Chapter 23, particularly Box 23.5 which indicates the nature of the procedure.

Sustainable and Acceptable Developments 14

Aim

To consider what a compliant building should be like

Objectives

* **to describe the principal concerns that society has for developments**
* **to analyse the concept of sustainability for buildings**
* **to explain the importance of each body of official powers affecting development**
* **to flag up future possibilities for developments**

Introduction

Since the 19th century professional practice has grown under a mass of accumulated legislation to ensure that buildings are acceptable. Controls over building, public health infrastructure, planning, environment and heritage have stepped together out of the urban slums and industrial chaos of the early times.

Ethos to acceptability

Many regard and talk about any control of a development or a building by public bodies as red tape, nanny state-ism, dead hand-ism' political popular-ism or interference. However, the controls are imposed because they are about reducing the number of deaths, illness and accidents, as well as increasing amenity, the quality of the environment, the quality of living and working space and making leisure and recreation more satisfying. Of course, problems arise when two or more controls are faced by a developer or some other persons and there is conflict between two or more officials — the outcomes or answers are sometimes "no" or there are less than satisfactory compromises. There may be an avenue of appeal, but the adjudication may not be satisfactory. This official ethos is the background to much that follows in this chapter.

Today, developments are "acceptable" when they are deemed so from the following professional perspectives:

- acceptable to the norms of planning control
- acceptable to the norms of public health and safety control
- acceptable to the norms of building control
- acceptable to the norms of environmental control
- acceptable to the norms of heritage control.

A further acceptability lies in the "buildability" of a proposed development once the other norms have been covered. Also, for a private developer keen to complete a development successfully, issues such as the following are important:

- the time taken to get though the controls
- the cost of implementing the development subject to such controls
- the impact in the market place of the quality to price ratio.

Acceptable is emphasised in quotes because none of the controls should be regarded as absolute when overriding or conflicting features in a development are presented — they become a norm or are a compromise between norms. In a sense the creation of norms and their compromise is part of the excitement of development in and of society; because society is now pushing at the boundaries of the existing norms with new norms. Perhaps a new professionalism is being created — "sustainability" control which will lead to sustainable developments.

Of course, change never stops, for instance, the Sustainable and Secure Buildings Act 2004 seemingly extends the building regulations into such matters as crime prevention, sustainable development and even composting (see below).

Sustainability and buildings

In recent years the government's thrust on sustainability is of considerable importance to the design process (and construction process). Sustainable property has many, if not all, of the characteristics given in Box 14.1. Whereas it is possible to design and build a new building to sustainable standards from scratch, it will be less easy to improve an existing building. Bolt-on sustainability will probably not cover all characteristics, cost more and have more compromises.

The principles of sustainability are, of course, embedded by the designer and executed by the builder. Compared with recent traditional construction, sustainability may increase the capital spend but the occupier should save on annual outlays. The user (and buyer) may use the checklist for assurance that the following features of sustainability, at least, have been considered during the both the design, budget and construction stages of the development process:

- locality, topography and orientation
- natural phenomena
- materials and components
- internal and external configuration and layout
- management of water

Box 14.1 Characteristics of a sustainable building

Locality	• transportation	• facilities provide for cycle parking and changing
Topography	• levels	• underground rooms are built to utilise changes in level
Orientation	• direction	• avoid frost valleys or "pits"
Natural light	• sunlight	• sunlight is utilised for ambience and for heating • use solar pipes for underground and enclosed accommodation
Natural water	• rainwater	• rainwater is utilised for gardening and, after treatment, for consumption and cleaning
	• snow and ice	• create icehouse • ensure roof loading is adequate for the locality
Materials and components	• sources • embodied energy	• sustainable sourcing is used • use of recycled materials • energy management recognises embodied energy
Configuration and layout	• movements • separation	• movements are minimised, eg of goods, persons, etc • vehicles are separated from pedestrian and cyclists
Water	• flooding • consumption	• designs eliminate or mitigate flood damage • automatic switch off taps used • metered supplies to induce reduction of consumption
	• grey water	• recycled from sinks and baths to treatment facilities for re-use in lavatories and gardens
Energy	• renewable source	• install facilities • link to the national grid to sell surplus electricity
	• conservation	• conserve fuel and power, eg double glazing, heat exchangers
Landscape or garden design	• structures • planting • water management	• install composting facilities • use plants for security, screening and shade • follow principles of xeriscaping, including drought resistant plants
Security	• systems	• access control • alarms, CCTV, and other intruder detection devices • guard geese and guard dogs • introduce gravel paths
Garden maintenance	• maintenance	• use grey water (see above) • store rainwater

- waste management
- energy management
- if appropriate, garden or landscape design and maintenance.

Planning and design control

A development is regarded as acceptable by the local planning authority when it receives planning permission. If an application is rejected by the LPA it may, on appeal, receive consent or a refusal from the Office of the Deputy Prime Minister (ODPM).

Planning legislation provides local planning authorities with a range of functions to control the way in which development is carried out. The scope is very wide and the functions are briefly indicated in Box 14.2. The Planning and Compulsory Purchase Act 2004 has overhauled the planning system with the aim of speeding up the planning process. A transition phase has commenced during which participants in planning will need to develop an understanding of the changes which are being introduced over the next few years.

Box 14.2 Planning authority's functions, in brief

Forward planning	• ODPM's PPS12 : *Local Development Frameworks* (LDFs) — provides guidance on forms of LDFs replacing structure plans etc • regeneration and other initiatives, eg the Plan for Sustainable Communities
Designation of areas	• various kinds of designation, including areas of outstanding natural beauty, conservation areas and simplified planning zones • they each have particular controls or characteristics
Public participation	• participation by the public is embodied in the 2004 Act's planning regime. • Community Involvement Statement (CIS) should demonstrate how the public have been involved
Implementation	• use of compulsory purchase orders for land assembly or key sites • use of planning agreements • use of partnership or joint venture arrangements
Development control	• decide planning applications and impose conditions, if necessary • planning agreements — planning obligations and contributions • modification or revocation orders • discontinuance orders
Special controls	• listed building regime — consents, refusals and compulsory purchase orders (CPOs) • conservation areas • tree preservation order regime • important hedges regime
Enforcement	• stop notices • enforcement notices
Compulsory purchase	• requirement for planning purposes • protection of listed buildings
Consultation	• for a development, conduct consultations with the relevant statutory consultee

---

Individual projects which do not have planning permission normally require planning permission unless the proposal is within one of the following:

- minor works not constituting development and, therefore, not requiring planning permission
- development within the General Permitted Development Order
- change of use within a class of the Use Classes Order
- development in a simplified planning zone category or that of an urban development corporation.

Generally, the district level councils are the development control and enforcement authorities. However, in Chapter 13, Box 13.2 shows in broad terms how the functions of town and country planning are divided between the various national, regional and local government bodies.

Building control or compliant construction

A number of statutes should ensure that buildings are compliant. Thus, standards of construction for buildings are provided for by the Building Act 1984 and laid down in the Building Regulations 2000 (as amended). The building control office of the local authority administers the way in which buildings are constructed. Box 14.3 sets out the parts of the building regulations.

Box 14.3 The scope of the building regulations

Part	Scope
A	• structural stability
B	• fire
C	• materials, workmanship, site preparation and moisture exclusion
D	• toxic substances
E	• sound insulation
F	• ventilation
G	• hygiene
H	• drainage and waste disposal
J	• combustion appliances and fuel storage systems
K	• protection from falling, collision and impact
L	• conservation of fuel and power
M	• access facilities for disabled people
N	• glazing
P	• electricity (a new Part)

Approved documents

Just fewer than 20 approved documents have been published by the ODPM and other bodies as non-mandatory technical guidance on the application of the building regulations. Although they are not mandatory their observance would probably be supportive of an architect or designer in refuting a client's allegation of a breach of contract.

The Sustainable and Secure Buildings Act 2004 develops the function of the building regulations to emphasise or introduce new areas of control. All the regulations are dealt with by building control officers or private approved building inspectors who carry out the day-to-day work under the regime. However, the use of a building may influence the application of the building regulations. Accordingly, certain public buildings are exempt buildings, eg schools.

In addition to the building regulations, there are over 40 Acts which apply additional regulations in particular counties, towns and other areas. Some aspects of various Acts are redundant and the Office of the Deputy Prime Minister proposes to repeal other aspects.

Energy management

Although the building regulations deal with energy and the like in buildings (Part L), this section takes a somewhat wider context. What is happening in the UK energy industry has immediate effect on the prospect of changes in the way buildings are designed and constructed.

Thus, it could be argued that the property industry will be affected by the security of the supply of energy from imported gas. This is being addressed by the following means:

- the storage of gas in developed underground reservoirs (see Chapter 1)
- the development of renewable energies.

These new energies are already resulting in demand for new services, and buildings or facilities in buildings, such as:

- small scale power generators in weirs in rivers
- increased efficiency in the use of energy by insulation of buildings and the use of heat exchangers
- installation and use of energy efficient white appliances in the home (particularly grades A to C)
- the development of on-site renewable energy generation systems, eg electricity generators and fuel pumps, for both domestic users and small businesses.

The management of energy is frequently perceived in terms of providing and maintaining cost-effective energy usage in buildings. However, a wider interpretation of energy management could include the movement of people and vehicles between buildings on a site or in a settlement. Layout design to ensure a reduction of vehicular movements should lead to substantial energy savings.

Disabilities

The Disability Discrimination Act 1995 provides that from 1 October 2004, the owners or lessees of premises and facilities providing services, eg playgrounds, shops, may be required to reasonably adapt them so as to enable disabled persons to conveniently, for example, shop or play in the property. Of course, in a given case it may not be possible to comply with the regulations, but this should be justifiable.

The Disability Rights Commission (DRC) was created by and charged under the Disability Rights Commission Act 1999 with overseeing good practice under the 1995 Act, including the arrangements for access to premisess. Although enforcement powers are available to the DRC, it seems likely that early practice will be implementation by persuasion, advice and information services and the like.

Where a business has not addressed any concerns about its premise, members of the National Register of Access Consultants (NRAC) or property professionals may give advice on the need for works to premises and other places.

Fire certification

Although Part B of the building regulations covers fire other enactments which require compliance impinge within buildings include:

* the Fire Precautions Act 1971
* the Fire Precautions (Workplace) Regulations 1997 (SI 1997 No 1840).

A development where a change of use is proposed may not involve building works, eg conversion works. The developer may proceed with owner occupation or a letting after obtaining planning permission or by virtue of the Use Classes Order. However, care must be taken to observe the above mentioned enactments.

The 1971 Act concerns means of escape from existing buildings where a fire certificate is required from the fire brigade, including:

* hotels or boarding houses — of certain internal arrangements of bedrooms
* factories, offices, shops and railway premises — over certain limits of employees and where they work
* factories — storing or using explosive and highly inflammable materials.

The local fire brigade will inspect the premises for the purposes of fire certification, specifying any necessary works to be carried out before the fire certificate is issued.

The 1997 Regulations require a degree of management by the occupier. For new build, compliance with the building regulations fulfil many of the physical requirements of workplace fire precautions, such as fire alarms, fire detectors and fire fighting equipment. Again, a change of use without building works should not prevent compliance with the regulations. For the owner occupier developer or landlord developer it is essential to comply with such matters as:

* readily available and systems for maintaining fire fighting equipment
* maintaining escape routes and exits
* trained staff to deal with aspects of policy regarding fire safety
* appropriate arrangements, including communication equipment to contact the fire and other emergency services.

Environmental controls

The Environment Agency and other bodies are largely responsible for environmental controls. The controls include:

* the requirement for an environmental impact assessment to be undertaken by a developer for specified developments (see Chapter 13)

- regulation, prevention and remediation of pollution of land and water resources
- statutory planning consultation on developments involving a cemetery, fish farming, refuse and waste depositing, mining, refining mineral oils and the like, storing mineral oils and the like, works to a stream or river's banks or bed and, finally certain developments involving operations to sewage, slurry, sludge and trade waste.

Environmental impact assessment

Developments are divided into two categories for the environmental impact assessments, ie Schedule 1 and Schedule 2 developments under the Town and Country Planning (Assessment of Environmental Effects) Regulations 1988 (SI 1988 No 1199). A mandatory requirement is imposed on developments in schedule 1. The latter is discretionary at the behest of the local planning authority. There may be some developments which, although not requiring planning permission, require an environmental impact assessment.

Radon protection

Radon is a natural radioactive gas which can cause illness, eg lung cancer. There are areas where it seeps from the substrata into the atmosphere and into cavities or underground accommodation in buildings. Part C of the building regulations requires that buildings at risk of contamination are constructed in a way which mitigates or removes the impact of the gas which has no smell and is colourless.

The local building control office will advise of the need for and level of protection. The areas where buildings require protection are to be found in many parts of the country including:

- Cornwall
- Cumbria
- Derbyshire
- Devon
- Leicestershire
- Lincolnshire
- Northamptonshire
- Northumberland
- Oxfordshire
- Rutland
- Somerset
- Staffordshire
- Wales
- West Yorkshire.

Asbestos management

Historically, asbestos was used in the construction of buildings. It is a highly dangerous substance — causing cancers, after many years after exposure to dust caused by disturbing it. There are many places where it was installed in buildings prior to the banning of it in construction in 1999. They include:

- lagging for pipes
- roofing slabs
- insulation for boilers
- damp proof courses.

The Control of Asbestos at Work Regulations 2002 (SI 2002 No 2675) imposes on duty holders (of non-domestic premise) a duty to manage asbestos. Briefly, from 21 May 2004 a management system must be created to:

- survey and record the position and types of asbestos in the premises (a register)
- create and maintain an Asbestos Register for the premises
- create a management plan to deal with each type of asbestos
- take action to deal with the asbestos in accord with the plan
- have a warning system in place so that maintenance and other workers know of the presence of any asbestos which may be disturbed by their work
- monitor and record the operation of the plan.

Heritage protection

Responsibility for the built heritage is given to a number of bodies as shown in Box 14.4. In some instances the bodies are regulatory and in others they act as owners or custodians.

Box 14.4 Some heritage organisations and their scope

Department of Culture, Media and Sport	• a statutory consultee for developments near Windsor Castle, Great Park and Home Park (3 km) and other royal palaces and parks (800 m) — affects on amenity and security
English Heritage	• a statutory consultee regarding conservation areas, ancient monuments and applications for listed building consent • maintains Lists of Historic Gardens and Parks and of Historic Battlefields • due to take over the day-to-day operations for listing buildings (this will require legislation)
Historic Houses Association	• owners of historic houses comprise the membership • promotes the houses and develops services to members on stewardship, taxation and so on
National Trust	• owner of a substantial heritage estate comprising buildings, coastal areas, gardens, etc
Georgian Society	• pressure or interest group on Georgian buildings and cultural history • informed of prospective demolition of buildings
Victorian Society	• see Georgian Society
Ancient Monuments Society	• see Georgian Society

Materials and components

Increasingly, designers, specifiers and contractors are required to give attention to materials and components for buildings and infrastructure in a way which focuses the likes of the following:

- using sources of supply which are managed in a sustainable manner
- embodied energy (see below)
- installing facilities for renewable energy from off-site sources or generation on site
- recycling of construction materials or materials made from recycled construction waste.

Embodied energy

Embodied energy is the energy consumed in the manufacture, storage and transportation of a component or material to the end user. In energy management, a comparison may be needed of any energy usage saving and the embodied energy of, say, a particular component.

Nature protection

Nature protection is effected by such bodies as English Nature. A number of international, national and local designations are used to highlight concerns and their importance (see marine nature reserves in Chapter 1). Flora and fauna as well as geological and other hard features are protected in different ways, but mainly by the designations. Several statutes afford protection to particular creatures, such as badgers and bats.

Emergency, health, rescue and safety

Many bodies control health and safety in development work as shown in Box 14.5. The scope portrayed extends beyond direct involvement with building, such as building control, the many organisations which impinge on the everyday lives of occupiers and owners of property. In summary, they are all directly concerned with making settlements or individual properties safe or worth living in.

Future

Innovation in development involves first-time work by inventors, owners and occupiers wanting to try new gadgets, etc and developers willing to test the market, either as investors or dealers.

Some items and developments described below have become standard for some market segments.

Early installations tend to be prototype or one-off life-style innovations of the inventor. There follows market research leading to testing, manufacture and marketing result in consumers needing and buying — for a few of the original prototypes.

Box 14.5 Scope of organisations concerned with health, safety and other matters

Ambulance service	• paramedic services at accidents and emergencies
Building control	• regulation and enforcement of building regulations (see above)
Environment Agency	• regulation and control of environmental standards (see above)
Fire service	• advice on fire prevention
	• fire certification (see above)
	• consulted on Fire Precautions Act 1971 and other legislation for works to buildings by building control officers
Gas industry	• regulation under the Gas Safety (Installation and Use) Regulations 1998
	• work by CORGI installer, ie Council for Registered Gas Installers
Health and Safety Executive	• regulation and enforcement of health and safety enactments, eg the Health and Safety at Work etc Act 1974
Highway authority	• standards set by the Highways Act 1980 covering skips, building materials, structures and roadside scaffolding
Housing authority	• standards set by the Housing Act 1985 for new and existing dwellings where works are to be undertaken
	• powers to order closure or demolition of unsafe dwellings
Military authorities	• afford protection in the event of emergency, eg fire service and BSE
National Health Service	• regulation and enforcement of clinical health and safety standards in hospitals
Police	• advice on crime prevention
	• attendance at accident and crime scenes
Water industry	• standards set by the Water Industry Act 1991 and the building regulations
Waste collection authorities	• Environmental Protection Act 1990 requires collection of waste from households, commercial waste (on request) and, perhaps, industry (waste disposal authority consent)
	• a new landfill directive requires separate disposal of no-hazardous and hazardous waste and, seemingly, reduces the number of sites that can accept the latter very severely

Listed buildings

English Heritage may be due to take over the day-to-day running of the operations of the control of buildings of special architectural or historic interest, eg listing. The Secretary of State will retain responsibility for overall policy. However, the proposed change will require legislation in a crowded parliamentary programme.

Part 5

Development Process

Property Development Process

15

Aim

To examine the "stages" of the development process

Objectives

- **to describe the roles and activities of those involved in development**
- **to examine the development process as a series of 11 stages, referring to other chapters**

Introduction

Whereas the last chapter looked at development from the perspective of the settlement, this chapter considers the development of an individual parcel of land. First, the chapter outlines the multifaceted activities carried out by numerous professionals and others. The principal function is usually construction but, second, it is appropriate to consider development as a number of stages, largely in sequence but overlapping in many respects. The providers are the developers described in Chapter 5

The so-called not-in-my-back-yard (NIMBY) feature of protest seems inevitable but it is one which a prospective developer will need to address. Acceptable development has many dimensions — some official, others which need to address the concerns of the NIMBYs and others.

Roles and activities

Each development is a complex mix of roles and activities which will differ in configuration from the next. The size of the project will have an important bearing on the number of different professionals, contractors or trades and others likely to be involved. In some instances there may be a forward sale or a pre-letting so the buyer or the leaseholder (or both) may have representatives in a monitoring role. Finally, if the leaseholder is the developer, the land owner will have the supervisory role. Boxes 15.1 and 15.2 are general lists of those participating in a project. The first box gives private sector participants and the second gives those in the public sector.

Box 15.1 Private sector roles and activities in the development process

Employer/client	• specifies the required new building
	• undertakes the developer role
(Buyer/leaseholder)	• a pre-let tenant or a forward sale buyer will add to the specification
Project manager	• overall management role for the development or the construction management role
Designer/architect	• designs the building and may manage the construction
Planning supervisor	• prepares the on-site health and safety plan
	• prepares the health and safety plan and manual for the future occupation, eg maintenance
Contractor	• main construction role on the site
	• appoints all or most of the sub-contractors
Sub-contractor(s)	• contractors for work packages — employer may nominate some sub-contractors
Quantity surveyor	• draws up contract for the work
	• prepares works documentation, eg bill of quantities or specification
	• prepares cost plan
Structural engineer	• specifies structure for the design
Building services engineer	• specifies building services for the design
Insurers	• arranges appropriate cover for the parties
Surety	• arranges property bond for the contractor (with the employer)
Interior designer	• designs and installs the interior décor and ornamentation
Landscape architect	• designs and supervises the grounds layout and planting
Gardener	• lays out the landscaper's design, ie structural works and planting

Box 15.2 Public sector roles and activities in the development process

Planning officer	• deals with informal applicant's enquiries
(local planning authority)	• receives and checks that planning application and fees comply with requirements
	• consults with outside organisations etc
	• considers application and consultations
	• recommends to committee (or if delegated, determines application)
	• notifies applicant of result
	• deals with appeal, if any
Building control officer	• deals with informal enquiries
Approved building	• receives application and plans
inspector	• decides on application and informs applicant
	• visits site works in accord with progress of work
	• signs off the completed work
Planning inspector	• receives appeal
	• hears representations
	• makes decision (unless minister recovers the application)
	• minister makes the decision (on major project) on recovered applications
	• informs parties of the decision
Statutory and	• receives details from the LPA
other consultees	• considers details in accord with own policies, etc
	• makes representations, eg the highways authority
	• if appropriate, gives evidence at appeal

Development process

Box 15.3 shows the main stages which must be done to finish a proposal for a development. It is necessarily brief and schematic, for instance:

- the stages may overlap to a large degree
- some proposals will not have the sequence described below.

Some of the stages are dealt with in more detail in other chapters as indicated in the box.

Box 15.3 Stages in the development process

Stage	Description of the stage	Chapter
1	• idea for a development	
2	• research and marketing mix	5
3	• site appraisal and selection	16
4	• development choices and decision	17
5	• planning, building and other controls	14
6	• land assembly and site acquisition	12
7	• site clearance and demolition	18
8	• design	15
9	• procurement of construction	19
10	• handover and commissioning	20
11	• promotion and disposals	5

Stage 1 — idea for a development

Actual development always follows an idea. Ideas are generated in the minds of individuals or those in organisations. The ideas come from numerous possible sources. Examples include:

- fashion — the buy-to-let market, developer improves property and lets it
- the receipt of a gift of a site — leading to thoughts of what to do with it
- legislation — the current prospect of a growth in casino developments follows the government's proposed liberalising of the gaming laws
- increasing local population — resulting in a local education authority building new schools.

Stage 2 — research and marketing mix

After the idea, the developer must quantify a need that the development will be saleable or lettable. The developer must firm up the idea by researching the prospective market to:

- confirm the existence of the need in terms of what it really constitutes
- work out what kind of development will satisfy the need
- make sure that the development's value, cost and quality ensure it is both physically feasible and financially viable
- find out whether the group is large enough to make the project profitable.

Thus, research will be required of the developer to sharpen the understanding of need which at the earliest stage is just the idea. A development does not meet merely one need but a number or bundle of complementary needs. Sometimes it will not be possible to satisfy all the needs because it would either not be physically feasible or financially viable.

Also, need for a property does not always mean that there is demand at a price. At some point in the research the developer will determine that there is a sufficiently large enough group (a targeted segment) available to buy or rent the proposed development, ie there is a market segment. Market segmentation is the output from such a study; as explored in Chapter 5 where a case study is given to illustrate the idea of marketing mix. (It should be borne in mind that the process is rarely as rigorous as Chapter 5 might suggest but developers will, hopefully, have "nous" for what their market wants.)

Stage 3 — site appraisal and selection

If the developer has a site, the selection has already been made and the development or the building component of the marketing mix will be "fitted" to the site. In a sense this always happens since the nature of the site will usually result in compromises on detail. However, house-builders buy land which will suit their purposes for immediate development if they do not have a landbank. Where they have a landbank which may have taken years to create, the research for each site coming on stream will extend to finding out the kind of development that the planning authority will allow.

An identified site will therefore have to be evaluated before development can be started. Chapter 16 explores the appraisal of an area and of a site in some detail and Chapter 24 covers valuations for development.

Stage 4 — development choices and decision

Financial appraisals are rarely, if ever, done only once. The prudent developer will know what outcomes are to be measured by a certain kind of appraisal, undertaking the different kinds of appraisals on the many occasions when the need arises. Chapter 23 is an examination of financial appraisal and the development budget.

Decision hierarchy

A developer needs to appraise any site according to the place it has in a hierarchy of decisions. Essentially, opportunities range from do nothing to redevelopment. There is a hierarchy of development proposals which depend on the nature or quality of the site. The developer has the opportunity to do something with the existing building or to demolish and start again.

Development equation

The developer usually as a good insight as to what can be done with a property but where there is uncertainty or a decision to develop may rest on fine margins the development inequalities portrayed in Chapter 23 may give insights to the answer.

Building quality

A proposed building must be acceptable to the authorities, as outlined in Chapter 14. At the same time in must meet the requirements of the occupier in terms of value, cost or quality (or all three). The value may be appraised as described in Chapter 24 and cost ascertained by a quantity surveyor. Capital cost is not necessarily sufficient information since the user may want to measure cost in various ways, including:

- annual running expenses — repairs, maintenance, insurances, rent (or the cost of borrowing to build the property) and other expenses of the building as such
- cost-in-use or life cycle costs — the present value of the steam of annual expenses of repairs, maintenance, insurances and so on, together with future renewals of the plant and machinery and the fabric of the building which wears out.

There are a variety of approaches to these items, eg to include inflation or taxation or both.

Stage 5 — planning, building and other controls

A seemingly chicken and egg problem seems to have arisen with this stage in that the designer's role requires the obtaining of planning and other consents. In fact this stage may be regarded as a "preliminary informal enquiries" approach to the regulatory authorities to ascertain what controls, conditions and standards might be expected. It should be of use to the valuer and estates surveyor in valuing the site for development purposes and in obtaining an outline planning permission, if necessary.

Of course, some or all of the information may have been obtained by the research (Stage 2) or by the site appraiser (Stage 3) but it is possible that the information is dated, has not been passed on or was obtained by another professional advisor.

Stage 6 — land assembly and site acquisition
Buying property for development

In readiness for development, the prospective developer has several legal devices which may be used to buy a site, particularly where the building has several ownerships, including:

- options — ie the right to buy or not buy a property within a particular period
- conditional contract, eg buying subject to planning permission being obtained by the developer
- buying with vacant possession
- buying through nominee companies.

The option enables the developer to effectively assemble the site without excessive expenditure but not completing until satisfied that a scheme could go ahead. Similarly, a conditional contract enables the developer to get out of it if the condition is not satisfied. Thus, if planning permission does not exist, the developer is not committed to buy until it is obtained.

Vacant possession means that no delays occur because the site has premises which are occupied by one or more tenants who have unexpired portions of leases or who want seemingly excessive payments for the surrender of their leases. Buying with vacant possession puts the onus and cost on the site owner-seller. In that way there should be no delays due to possession being retained, such that the developer is holding the site and has paid for it.

Finally, in the past, for large sites with many properties in different ownerships, developers have bought secretly over long periods, using several nominee companies. Generally, this approach was used in the hope that the prices paid reflected the existing uses of the properties rather than any marriage value.

Buying and selling land is covered in Chapter 12 and valuations in Chapter 24.

Development briefs

The seller of land may prepare a development brief for the site before it is offered on the market. Here, in the early stages, the developer is likely to be in competition with others before a finally approved scheme, based upon the development brief is agreed with the owner. A development brief (and a building lease or a building licence with an agreement for a lease) is usually required by the landowner when retaining adjoining land or where the land is part of a large dispersed estate.

Stage 7 — site clearance and demolition

The clearance and demolition of an area or a site in readiness for development or redevelopment is dealt with in more detail in Chapter 18. Here some concerns of managing the process are explored from the developer's perspective.

Possession from lessees

Where a developer is required to obtain possession from lessees, two approaches are used:

- where the leases are long and the developer does not want to wait, buy the lessees out at a mutually acceptable price
- use the power to seek a court order for possession which is available under the Landlord and Tenant Act 1954 (for business tenancies) or other relevant statutes.

To obtain a court order, the developer is required to show a good case by demonstrating several if not all of the following:

- planning permission has been obtained
- finance for the development is available
- plans and specifications for the building works have been drawn up
- tenders for the construction work are ready.

Archaeology, contamination and other features

The appraisal of the site or the informal planning and other enquiries should have provided information about special planning controls, archaeology and contamination and other matters. All of which will need to be considered prior to demolition. Box 15.4 gives an aide-memoire to some of the points which should be covered prior to demolition. Many of these concerns are reviewed in Chapter 16.

Box 15.4 Concerns which the developer needs to address prior to demolition

Tenant security procedures	• developer to have data for making a case for possession
	• put all relevant dates in the future or cautions file
Offers of rehousing	• common with public and voluntary sector projects
Offers of alternative business	• a ground for getting possession
premises	• adds to rental income if tenant takes new accommodation being developed
Planning consent	• comply with all notices
	• have data collected for the application
Listed building consent and	• follow procedures
ancient monument consent	• have data to make the case
Tree preservation consent	• follow procedures
	• have data to make the case
Important hedges	• follow procedures
	• have data to make the case
Highways diversion or	• follow procedures
closure procedures	• have data to make the case
Temporary lettings and licences	• ensure duration does not delay the start or other important issues
Pipes, cables and oversails	• make early approaches to utilities
— diversions and plan for	• prepare plan for the demolition contractor and others
demolition contractor	
Fencing and boundary contractor	• specify requirements
Insurance policies	• compulsory insurances in place
	• contractors and professionals have cover
	• own policies cover requirements
	• ensure those who have possession of the site or part have cover, eg archaeologists, prospective buyer or leaseholder
Archaeological survey	• consider use of the British Property Federation standard form of agreement for developers and archaeology groups
	• ensure any funds are available for the dig

Stage 8 — design

The final completed building is the result of a final design which is the designer's initial interpretation of the client's brief (once it has been established) and as changed during the design process, eg for large projects, to allow for the planning supervisor's health and safety requirements. However, it will usually be modified to account of such matters as:

- the limitations of the client's budget
- statutory regulations
- the contractor's interpretation of the working drawings
- the availability of specified materials and components
- the client's variations during construction.

(The issues touched upon here are amplified in Chapter 19 where there is a brief treatment of various procurement approaches.)

Stage 9 — procurement of construction

One of the many procurement approaches will be used by the client (employer) to obtain the building's construction in accord with the client's brief. Professional advisers will recommend an appropriate standard form of contract, eg JCT, having considered the project and the general concerns of the client. Chapter 19 compares the ways of arranging the construction of a building, ie procurement.

Stage 10 — handover and commissioning

The contractor hands over the building to the employer in a state which enables any fitting out and preparedness for operational activities (see Chapter 20).

Stage 11 — promotion and disposals

Sales and promotion of the property will not be needed where the employer is intending to occupy the property unless part of the premises is going to be sold or let. However, where the property is being sold by a dealer or is being let by an investor, a promotion and disposals programme will be required. The estate agent's role and activities are explored in Chapter 12 in a general way.

If the property is specialist, a shopping centre, for instance, the kind of approach examined in Chapter 3 will be needed to ensure that the optimum tenant mix is achieved. Similarly, a residential development requiring a multi-letting approach may need the specialist letting management of a firm of letting agents, eg one who belongs to the National Approved Letting Service.

Site Appraisal and Selection

16

Aim

To establish the criteria for the successful development of a site

Objectives

- **to identify the roles and activities in site appraisal (and refer to other chapters)**
- **to show how the developer's likely main concerns are addressed on each topic**
- **to identify other concerns (and refer to other chapters)**

Introduction

The successful development or redevelopment of a site comprising land and buildings requires the resolution of many problems. The solution to the problems lies in relating specific attributes of the proposal to identified site selection criteria. Briefly, the following embrace the many criteria which need to be considered:

- physical condition
- planning history and context
- legal status and context
- financial resources.

The content of this chapter is grouped under these headings; but each is further sub-divided. Essentially, the perspective is given as a series of checklists under each heading. Bearing in mind that some items in a checklist will not necessarily be relevant, each item should be ticked as covered when a property is being considered for improvement, redevelopment or development. (Of course, some of the topics may also be relevant when property is being purchased for owner occupation or letting.)

Many of the topics in Chapters 2, 3 and 4 are germane to investigations for purchase or development of property. They relate to the occupation and use of the property — generally, as evidenced by the existing legal documentation: whereas this chapter looks at the property and its environment from a different practical perspective, namely:

Can its proposed new state (after building works or change of use) be successfully brought about? If not, what development, if any, can be done on the property?

Roles and activities

Where a role is identified but not dealt with in others chapters, it is covered below. Otherwise, the reader is referred to other chapters where roles and activities are covered in some detail.

Surveys and studies

A survey or study of a site will almost certainly cover one or more investigations, perhaps involving more than one specialist. Each expert will undertake a study, which will probably comprise a desk research and an on-site appraisal. The surveys likely to be included are:

- for unregistered land, possibly a deed search and local searches
- a valuation survey
- a structural survey
- a geomatic or land survey
- an archaeological study
- prior to demolition, a pipes and cables survey
- where previous contaminative uses are known or suspected, a contamination survey.

Physical condition

Many aspects of the physical condition result in the prohibition or limitation of development or in high cost to remediate them. In particular, the following are current thresholds for many sites:

- for brown field sites, severe dereliction of the site and of land in the neighbourhood
- known or suspected previous contaminative uses
- history of flooding in the locality
- known or suspected evidence of important archaeological history
- lack of infrastructure in the locality.

Levels and contours

On the one hand, the shape and slopes of land may lend themselves to attractive layouts of an estate and orientation of buildings. Opportunities to use any down slopes for drainage, water courses and the like should be considered by the designer. Similarly, underground rooms, garden features and gravity assisted movement may be suggested in some situations. On the other hand, landslide and landslip may feature on sloping land, requiring retaining walls and underpinning. Similarly, rainwater may run off a slope, causing local flooding. Storm ditches and barrier walls may be the answer. Finally, the bottom of slopes may harbour frost spots and trapped water where the gardener may experience problems with the water features and plants which are not hardy.

Flooding

The prospect of flooding is now well documented on the Internet by the Environment Agency. The degree of risk of a flood occurring can be pinpointed to the postcode for almost every building. This is an important aspect of insurance cover for flooding.

Natural features

Care needs to be taken with natural features on or near a site. If nature designations exist, any work must be preceded by the official permissions.

Solar energies

Generally, the orientation of buildings and gardens should be such that they embrace the sun for the joy it brings. The sun can be utilised in the design of buildings:

* to generate solar energy by convection for heating water
* by photovoltaic conversion for power, eg using special roof tiles
* by bending solar light in reflective pipes to bring it to dark interior accommodation
* by heat transfer pumps to warm interior parts of the building not touched by the sun.

Soil and substrata, including minerals

The gardener will be concerned about the topsoil, ie in terms of its condition and type, which together with water supply and sunlight will determine the kind of and abundance from the garden.

Subsidence, landslip and the like

Many areas are subject to subsidence and the like. A desk study and a site appraisal should include an evaluation on such matters as:

* coal mining and the extraction of other minerals
* dene holes
* landslip
* old landfill sites
* filled in ponds and underground streams and rivers
* coastal shoreline management — managed withdrawal or retreat.

Generally, a site survey may indicate evidence of such matters but it is more likely that local archives, newspapers and other material used for desk studies will be more forthcoming. Similarly, nearby residents may have memories of these matters.

Water

Subject to abstraction permission, the availability of clean natural water is a bonus in modern times. If sufficient for agriculture, home or industry the occupier-user may be assured of water. Incursions against the ownership rights to abstract from any rivers, streams and other waters should be checked on-site. Similarly, any rights to piers, wharfs and mooring rights should be safeguarded. On buying property, any unused capital allowances for walls and embankments may be available from the seller and should be checked.

Boundaries

The ownership of boundary walls, fences and gates should be checked and any obligation to repair and maintain them noted. To avoid or minimise the likelihood of a dispute arising, works to any party wall should be in accordance with the Party Wall etc. Act 1995.

Wind and rain

Prevailing winds, perhaps carrying rain or snow may determine the location and orientation of a proposed building and the positioning of outside drainage and water courses. In a garden, the careful placing of hard features, eg walls as windbreaks, and the judicious planting of trees and shrubs should ease the users' discomfort in the wind.

Frost and snow may cause danger. In high ground locations regular annual heavy snow deposits may require somewhat abnormally strong roofing support. In frosty places in the grounds of a building, planting requires hardy plants.

Waste and landfill

The presence of a nearby waste treatment facility or a landfill site may curtail the kind of development the local planning authority would permit or, indeed, what the developer might expect to sell or let at a price to make the development worthwhile

Planning history and context

An examination of the planning aspects of the site is a prerequisite to determine any need for an application for planning permission to change the use of the property or carry out building works. The following principal matters will need to be considered:

- the planning history of the site
- existing use rights
- the possible application of special planning controls by the local planning authority
- extant or proposed planning policies for the area.

Planning history

Possible sources of information about the planning history of the site include:

- the estate office records
- Ordnance Survey plans
- the local land charges register
- the planning register at the local planning authority office.

The estate's records or the local planning authority office or both are likely to contain information on or copies of matters which may give insights on the way in which the local planning authority may handle a new application, including:

- any previous planning applications and their determinations
- any appeals made and their determinations
- any s 106 agreements (on planning obligations)
- revocation or modification orders on a previous planning permission
- discontinuance order affecting a previous use of the site with, perhaps an alternative planning permission
- any Directions affecting the local application of the General Permitted Development Order or the Use Classes Order.

If any planning permission appears to be operable the local planning authority should be able to confirm that it has not expired and is, therefore, valid.

Special controls

For local planning authorities a number of development controls, designated "special controls" enhance their capability to influence or prevent development of a site, including:

- listed building consents
- tree preservation orders
- revocation and modification orders
- discontinuance orders
- compensation payments
- stop notices.

Details of past events involving these, if any, should be available in the planning history of the land or in the land charges register.

Planning permission

A developer must obtain planning permission for any works or change of use (basically development) unless the proposal is one of the following situations:

- the works are not development within the meaning of the Town and Country Planning Act 1990
- the proposal is permitted development in accord with the General Permitted Development Order
- the development is a change of use within a use class of the Use Classes Order 1987
- a valid planning permission already exists for the proposal
- the proposal meets the conditions for a deemed planning permission within a Simplified Planning Zone or one of the remaining Enterprise Zone.

Details of this kind of information will be at the planning office, eg development plans, planning development frameworks, or in the planning register's details for the site of the local planning authority.

Use Classes Order

Generally, a developer may change from one use to another without the need to obtain planning permission provided both uses are in the same use class of the Use Classes Order. There are currently 11 use classes which are grouped into four Parts, A to D, of the Order. From time to time the configuration of uses is changed. Also, some uses do not fit into a class — they are *sui generis*, eg haulage use.

General Permitted Development Order

The GPDO grants planning permission for development specified in its schedules without the need to make an express planning application. A large number of developers, if not all, fall within the ambit of the GPDO, but most of the developments are of a minor nature. Although a planning application is not required in a few instance there may be a requirement for an environmental impact assessment to be undertaken.

Legal status and context

Land tenures — feasibility

A development must last long enough to be worthwhile or pay for itself. If the proposal is on leasehold property and period of tenure is too short, eg a lease of five years, there would not be sufficient time to amortise the development costs of a high rise office block — a lease of, say, 99 years would seem to be more appropriate.

Vacant possession and occupations

When the seller of land is an owner occupier the developer normally obtains vacant possession on completion of the sale. Indeed, the prudent developer will visit the site on the day of completion and inspect the property to ensure that vacant possession is to be given. If vacant possession is not being given, a call to the solicitor will delay the completion until compliance is achieved, particularly where time is not essential to the development programme.

However, when the site is occupied by leaseholders, the freeholder may be selling without offering vacant possession, ie the sale is subject to the leases. Here the developer must either:

- wait for each lease to end and take the properties in turn, or
- buy out those lessees who are willing to release their property before the day on which the lease ends.

When business lessees are bought out they may want to negotiate for marriage value, ie seeking some of the development value. If the lessees are unwilling to surrender, the developer will have to obtain possession at the end of the lease using rights under the Landlord and Tenant Act 1954. Where other kinds of tenant are in occupation and protected under other legislation the developer will need to seek possession under those statutes.

Restrictive covenants

A bar on development is frequently found in the form of a restrictive covenant. For instance consider a site subject to a restrictive covenant prohibiting the sale of alcoholic beverages. A developer wanting to build a supermarket would be reluctant to consider it because a prospective operator of the supermarket would normally want an off licence. Ways of discharging or modifying restrictive covenants are considered in Chapter 17.

Use of compulsory purchase

Local planning authorities do have powers to use compulsory purchase to enable the implementation of their development plans for the area.

Access to neighbour's land

An owner, who needs access to a neighbour's land for essential repairs and maintenance to the property, may seek an order from the court. If granted the entry is lawful and the work may be carried out. Any damage must be made good or compensation paid to the neighbour.

Unauthorised possessions

Title and land registration and other desk investigations will confirm ownership of the property or estate but an inspection on the ground will usually be needed to ascertain:

- squatters and other occupations, such that adverse possession might be claimed
- trespassers
- possible dangers to authorised visitors and others, such as derelict or damaged buildings, fences and walls.

Authorised but casual possessions

In some instances there may be tenants or licensees who, although authorised to be on the land by the owner, may be there on a casual basis. Care may be needed to establish the exact nature of their rights and how best to deal with them.

Authorised but formal possessions

Formal tenancies and licences would usually be picked up by the investigations handled by the solicitor. Where they subsist on site, details may be confirmed or obtained for the first time by interviewing the occupants. Concerns for the developer are that:

*   delays in obtaining possession will hold up the project
*   unbudgeted costs may be incurred in obtaining possession.

Financial resources and context

Planning contribution or planning gain

Hitherto, the planning system has provided for planning obligations under s 106 of the Town and Country Planning Act 1990. The agreements between a developer and the local planning authority could be payments in cash or in kind. Thus, in any assessment of the site, the estimated cost of any works or any proposed payment needs to be taken into account in the financial appraisals or budget.

In future, under the Planning and Compulsory Purchase Act 2004 the developer may be faced with a choice. The local planning authority is empowered to draw up a schedule of planning contributions for particular developments which the developer may choose to pay. Alternatively, the developer may choose negotiate a payment or carry out works instead of planning contribution. The latter is not yet in practice and is due to come into operation in 2005.

Grants and loans

Any grants and loans which are available from public or voluntary bodies need to be taken into the development budget.

Local taxation

Once a development is completed, the prospect of local taxation on the occupier (or owner in the case of vacant property) arises. Where different sites are being appraised the level and incidence of the local tax on alternative developments on each site may need to be considered. Where there is any prospect of a late occupation, the vacant property tax should be included in any risk assessment which accounts for delayed lettings. Similarly, the budget for the proposed development might allow for such voids.

Marriage value

When two interests in land are merged there is sometimes a release of value, called marriage value. For instance, when the leaseholder buys the freehold, the value of the unencumbered freehold is greater than the sum of the values of separate leasehold interest and the separate encumbered freehold.

Environment and amenity

Any developer will want to be assured of environmental concerns which may affect a proposed development. The website for Homecheck, *www.Homecheck.co.uk*, gives information on a number of these, namely:

- flood
- subsidence
- radon
- coal mining
- landfill
- historical land use
- pollution.

Various aspects of these are touched upon in this volume.

Air quality

Data on the quality of air in a locality is available. Pollution due to different agents is measured mechanically by local recording devices. It is shown on quantitive scales, but the seriousness of any pollution is expressed qualitatively.

Contamination

Ground pollution normally requires desk research and field activities. Where contamination is suspected, appropriate remediation is required before development can take place. Chapter 18 on site clearance and demolition looks at the ways in which evidence of contaminative uses will normally be revealed by a thorough desk study.

An on-site appraisal of a site which is thought to be contaminated will show up the pollution pathway between the source contaminant and the receptor. If contamination is confirmed by the appraisal, the owner will need to have commissioned or commission a report on what remedial work will be required for the proposed development.

High hedges

The Anti-social Behaviour Act 2003 offers those overshadowed by a neighbour's high hedge a means of relief. Various steps are specified in the Act which for this purpose came into effect on October 2004. If the neighbour is unable or unwilling to deal with the hedge, the aggrieved person may ask the local authority to act. An approach by the authority which is followed by inadequate action or inaction will enable the authority to enter the land and cut down the hedge to two metres, charging the owner for doing the work.

Radon

Radon is a natural occurring gas which may accumulate in buildings, particularly in basements where there are granite sub-strata. Although inert, it is cancer forming and any basement should either be air conditioned or sealed against the gas.

When buying property, the local building control office usually has information on radon treatment for buildings. Also, Internet databases may provide an initial assessment of the incidence of radon in a locality. Should its presence be suspected a survey should, prudently, be carried out and, if necessary, remedial work undertaken.

Heritage

"Heritage" covers many different aspects of land and buildings. In this section the topics covered are archaeology, monuments and historic buildings, cultural and historical land uses, buildings with architectural merit and protected areas. (The next section covers natural heritage.)

Archaeology

The period to some 10,000 years back may be regarded as the period of archaeological heritage in the British Isles. It spans the archaeology of the Stone Age, Bronze Age and Iron Age into the Roman era and on to the archaeology of the second World War — merely 60 years ago. Since then the bases of new archaeologies are being created with the demises of such lives as "quaint rural", "coal mining", "heavy industry" and "port or dockyard". Of course, hereafter the rise of reminiscence records, media technologies and tourism's industrial, open air and living museums will enable a kind of living archaeology.

Built heritage

Even so society protects its past with legislation or other means to prevent the spoliation, particularly for the following:

- listed buildings
- conservation areas
- historic battlefields
- historic gardens or parks
- ancient monuments
- archaeological remains.

Natural heritage

As with the built heritage, society protects it natural heritage with a host of site designations and other means, including:

- animals
- plants
- protected areas
- trees and important hedges.

Development Decisions

17

Aim

To explain the rationale of decision-making in development

Objectives

- **to identify and define the different types of value in property**
- **to explain the nature of profit in property transactions and works**
- **to consider the decision on whether to develop or not**
- **to examine finance and taxation in making decisions to development, referring to other chapters**

Introduction

Vacant property sometimes has potential for development. The owner may realise or at least think that the property could have something done to it or another person may have a clear idea about what could be done with it and makes an approach. The owner has several options:

- do nothing — nothing is likely to happen unless the other person is a body with powers of compulsory purchase or is a potential squatter
- mothball — minimum repair, redecoration and improvement, together with security (again compulsory purchase or squatters may loom)
- prepare for owner occupation or letting — repair, improve and redecorate to one's taste or to market requirements
- redevelop and occupy, sell or let
- sell or let as it stands or for redevelopment.

In general this chapter examines these options with the view to identifying key decisions nodes in the process of development (or otherwise).

Land values and development

A major political and social issue in society has been and still is the handling of both realised and unrealised land values. In brief, when a new settlement is mooted or an idea for a new road or railway is proposed, property owners can usually expect land values to be affected — some adversely others beneficially. Space does not allow more than a cursory review of the attempts garner betterment or allow for worsenment.

Existing use value and development value

The notion of development value is the key to understanding the ways in which governments have sought or seek to tax or otherwise gather increases in value due to development. Essentially, planning permission is seen to create development value (and any perceived prospect of obtaining planning permission "creates" hope value).

Obtaining planning permission for the development of a property usually, almost invariably, has a beneficial effect on its open market value.

Box 17.1 Existing use value and development value

Example: A property is a house with a large garden which it is expected could be sold in the open market for £200,000 as a house. However, with planning permission for several houses on the land, it could be expected to fetch £1,000,000 in the open market.

In technical terminology:	Open Market Value	=	£1,000,000
Less	Existing Use Value	=	£ 200,000
	Equals Development Value	=	£ 800,000

Numerous attempts to tax some or all of the £800,000 (or more accurately the development value) have been made since the end of the second World War — particularly in the following instances:

- development charge under the Town and Country Planning Act 1948 (1948 Act)
- existing use value compensation under the 1948 Act
- short term capital gains tax under the Finance Act 1962
- betterment levy under the Land Commission Act 1967
- development gains tax under the Finance Act 1974
- development land tax under the Development Land Tax Act 1976
- planning gain or planning obligation — under various Town and Country Planning Acts
- conceivably, planning contribution under the Planning and Compulsory Purchase Act 2004.

Generally, national taxation imposes tax on development value, particularly as follows:

- income tax and corporation tax catch any development value reflected in rents and premiums

- capital gains taxation is charged on development value in capital gains
- inheritance tax catches development value aggregated in the deceased's estate.

However, the latter taxes are not specifically related to the issue.

Risk and profit measurement

Entrepreneurial approach

For the entrepreneur the assessment of risk may be quantitative, but is often with a qualitative dimension, in the sense that it is hunch-induced or a gut feeling — in other words it is intuitive. Of course, the entrepreneur is knowledgeable and experienced about the market and is likely to be constantly consciously seeking and unconsciously absorbing information about

- what is happening in the market place, eg legislation
- trends in prices of property
- movements in costs of buildings, wages and materials
- customers' reactions to the quality of accommodation, eg in terms of buying or renting.

For a new building, in a residual valuation, risk and profit estimated as:

- say 10% to 15% of the value of the completed building (the gross development value), or
- 15% to 20% of the cost of the works.

A similar amount would be allowed in a developer's budget to assess the amount available for land purchase.

Contingency for the unforeseeable

However, for a refurbishment proposal of an old building, it is common to build into an appraisal an extra item for a contingency of, say, 5% of the cost of the works to allow for the prospect of unforeseen works, eg the removal of asbestos. Similarly, for brownfield sites, a contingency is allowed for unforeseen works due to hidden underground structures, eg foundations, old storage tanks. Where there might be the possibility of contamination of the land being valued, the valuer would normally build a contingency probability in the appraisals and, then, a final deduction to allow for "stigma". Of course, there will be estimates of the work needed for remediation of the land but the contingency would still be allowed. (Needless to say, contaminated land and buildings are extremely difficult to value!)

Effect of the process of planning permission

Land without planning permission and, seemingly, with no prospect or hope of planning permission, has an existing use value, eg a house as a house. Where something happens to create a "hope" of planning permission, the value will rise to *existing use value with hope value* (as an added increment). Prospective buyers make some kind of judgment and bid to pay the extra value — not a large amount where there is not a great hope. If the market has evidence of such transactions or of a trend in such transactions, a plot could, conceivably, be valued.

Once the process of certainty enters the market, ie that a plot will obtain planning permission in due course, the quantum of the allowance for risk and profit in successive valuations changes as the application progresses. This phenomenon was illustrated in valuations presented to the Lands Tribunal in a case concerning a modification order and discontinuance order, namely *Excel Markets Ltd v Gravesend Borough Council*. Thus, as the developer took steps towards:

- first planning permission
- second, building contract
- third, commencement of site by the contractor

so the risk and profit in appraisals at each stage decreased. Each successive notional sale of the developing site was valued at a higher price, the developer was successively reducing risks has each later stage was reached. The risk and profit in the first valuation was 20% and in the last 2.5%. The effect was to increase the compensation which Excel Markets Ltd obtained.

To develop or not?

At any time, the owner of a greenfield or clean site has choices; choices to do nothing, to sell, to let, to improve, or to redevelop. Each choice will require appraisals based upon objective open market transactions and subjective appraisals of such transactions taking the owner's financial, fiscal and other relevant circumstances into account.

As described in Chapter 16 many factors will affect the outcomes of a development decision-making process. For instance the cost of heavy pollution of the site may outweigh any increase in value due to the allowable proposed development.

Develop

Essentially, the decision to develop revolves about the prospect of bringing about a release of development value over and above the existing use value. A consideration of a few inequalities should reveal the impact of the variables in any development situation.

Inequalities

The main decisions in development are whether to develop land or whether to redevelop an existing property. Essentially, the present value (existing use value) (EUV) will be compared with the value of the land with the potential for development open market value (OMV) (or with a planning permission). If there are grants or taxation incentives available these may be taken into account. Discounted cash flow techniques may be used to work out the various values for the development.

Box 17.2 The development inequalities

Development is unlikely to take place where:

- EUV > OMV with PP
- EUV > OMV with PP + Grants + Capital Allowances

Development is likely to take place where:

- OMV with PP > EUV
- OMV with PP + Grants + Capital Allowances > EUV

EUV = Existing Use Value
OMV = Open Market Value
PP = Planning Permission

Leaseholds

Where leasehold property is to be redeveloped, the need to amortise capital over the duration of the lease needs to be considered; if the term of the lease is too short the works may not be worthwhile in terms of recovering the expenditure. For tenant's works which are not redevelopment, say relatively minor works which alter the property in some way but do not improve it, the tenant may even have to restore the property when the lease comes to an end. However, tenant's voluntary improvements may be eligible for compensation where the landlord obtains possession at the end of the lease. This will be the lesser of the cost of improvements or the increase in value of the landlord's reversion. (Compensation is not payable in every situation, eg where the landlord obtains possession for redevelopment.)

Refurbishment, makeover or extension

The decision to improve or do a makeover, has a similar logic to that given above.

Historic and other special property

A listed building owner who wishes to redevelop or modernise the property may face serious official opposition from the local planning authority. Apart from limited exceptions, there is a presumption against demolition and redevelopment of listed buildings, ie those buildings formally declared to be of special architectural or historic interest. Again, a refurbishment proposal to modernise such a building may result in a conflict of decisions between the planning and building control officials.

Finance and taxation

Finance

For the developer alternative sources of finance are likely to be available. Where the terms and conditions of the offers differ, appraisals may be needed to show the developer the best choice might be made. For instance, a developer may be offered, say, a mortgage by one finance company and funding by a deep discounted bond by another — the deep discounted bond may be offered on alternative arrangements! The appraisal chosen must be able to cope with the offers, see Chapter 25.

Taxation

Every choice of action by the owner and every funding arrangement are likely to have their own taxation implications — these too must be appraised. Status and the type of transaction will be important, but the type of building is important when capital allowances are being considered, see Chapters 26 to 30.

Site Clearance and Restoration 18

Aim

To present the problems and approaches to site clearance

Objectives

- **to identify the roles and activities in site clearance**
- **to describe the way property awaiting clearance is managed**
- **to outline the special treatment of contaminated land**
- **to show how government supports the clearance of dereliction and contamination**

Introduction

Clearance of an area or a site prior to the carrying out of a development proposal involves careful planning and execution of the likes of the following:

- obtaining possession from tenants and other authorised occupiers
- removal of any unauthorised occupiers
- removal of discarded chattels, plant and machinery and deleterious waste
- cutting off of services, such as gas, electricity, telecommunications and water
- removal or diversion of pipes and cables
- closing or diversion of roads, footpaths and other rights of way
- fencing or ditching of boundaries
- boarding of apertures in buildings
- remediation of contamination
- security, including guarding
- arranging archaeological survey, if any, in sensitive areas
- demolition of buildings and structures.

There are other areas where development works may be intended but they have special problems. Some of the activities mentioned above will be undertaken. However, the scale of clearance and restoration which will be needed is not likely to involve a private developer in the first instance. Special governmental involvement or intervention will be needed. Various agencies, controls or support will normally be needed to deal with them. The areas might include:

- abandoned coal mining, quarrying or opencast mining areas
- areas where quarrying or opencast mining is taking place
- sites where there are decommissioned nuclear power generation installations or waste — the so called nuclear legacy
- areas in economic and social decline
- towns or villages where a natural or man-made disaster has taken place
- areas where successful terrorist incidents have taken place.

Roles and activities

Box 18.1 shows who is involved in site clearance, and what they do.

Box 18.1 Professionals and others involved in the clearance of and demolition of buildings on a site (other than contaminated land — see Box 18.4)

Developer (landowner)	• appoints contractors • prepares pipes and cables plan
Highways authority	• close or diverts any roads
Local planning authority	• empowered to grant planning permission • deals with listed building or conservation area consents • deals with tree preservation orders, important hedges and the like
Valuer and estates surveyor	• secures possessions against occupiers • makes planning applications
Demolition contractor	• arranges demolitions • specialists deal with asbestos and the like (see Box 18.4 below for contaminated land)
Fencing contractor	• secure the site and buildings with fencing and other barriers

---

Property awaiting redevelopment

Redevelopment proposals for an area or a site will eventually result in a mixture comprising one or more of the following:

- vacant buildings or plots

- completely or partly let properties
- land or buildings subject to unauthorised occupation by squatters or others
- public roads, footpaths and the like
- pipes and cables over, on or under land
- burial grounds
- land or buildings which are contaminated or suggest contamination
- land for which there is archaeological evidence.

Management strategy

The owner will need to develop a management strategy which ensures that possession for redevelopment will be obtained in time for the handover to the contractor employed to carry out construction. Box 18.2 illustrates many of the matters which must be covered.

Box 18.2 Management of property awaiting redevelopment

Possession against tenants	legislation may be available	• compensation may become payable to the tenant for loss of security • tenant may be eligible for alternative accommodation
	negotiation may be the only approach to hand	• tenant may seek marriage value • negotiations may become protracted • may be appropriate to use agents and nominee principals
Fencing, ditching and boarding of apertures in buildings	carry out fencing as soon as land or buildings are cleared of occupiers	• fencing and boarding will deter entry by trespassers, squatters and others • ditching will deter vehicular entry to land by fly-tippers, travellers and others • boarding of doors and windows will deter entry by children and others
Painting	deterrent painting on walls, pipes and other surfaces	• anti-graffiti paint on walls and other flat surfaces may deter activity • anti-climbing paint on pipes and railings may prevent successful entry
Patrols and surveillance	caretakers and security guards	• on-property personnel are likely to prevent or detect unauthorised entry to otherwise vacant property • personnel should be licensed by the Security Industry Authority
	equipment, eg alarms and CCTV	• monitored detection and surveillance equipment should enable increased security

---

Occupied land and buildings

In managing occupied property care should be taken to ensure that decisions and actions do not prejudice the prospective handover date, eg leases should not be renewed without a full investigation. The legal procedures for obtaining possession are strictly observed and should not be compromised. It follows that for every lease or tenancy the case for obtaining possession should be carefully considered and the appropriate information obtained and action taken as necessary.

Vacant property

Vacant buildings which are not to be retained for refurbishment may be demolished or put into mothball maintenance until demolition. Short-term lettings may generate worthwhile income, eg advertisement hoardings; otherwise early demolition may be preferred — particularly if this would avoid liability for local taxation.

Clearance and demolition

Planning and demolition

The Planning and Compensation Act 1991 states that demolition requires planning permission, but there are many instances of permitted development under the Town and Country (General Permitted Development) Order 1995 (SI 1995 No 418). However, special consents are required in certain instances, including:

- demolition of a listed building
- demolition of a building in a conservation area
- certain ancient monuments.

Planning and advertisements

A developer may want to have advertisements erected on a site, perhaps for temporary or permanent income or for site security. The need for planning permission will require investigation. Some types of advertisement are, exempt or permitted without planning permission by virtue of the Town and Country Planning (Control of Advertisements) Regulations 1992 (SI 1992, No 666), as amended.

Security, fencing and boarding-up

Security and other contractors provide straight forward fencing, boarding-up and security services. A sensitive site may require special treatment with such items as:

- closed circuit television
- biometric alarm devices to detect the presence of humans or animals
- lighting and security lighting
- access to the site controls, eg biometric recognition devices
- guarding.

Diversion or removal of cables, pipes and oversails

Demolition of buildings and structures must be preceded by a careful check to locate the cabling and pipes of dormant services to the vacant buildings. However, the check should also identify active pipes and cables, including oversails, which burden the property, such as the wayleaves and easements of service undertakers or neighbours.

Prior to demolition the providers or owners of services, such as cables, pipe sand oversails, should be given the opportunity to:

- protect their apparatus
- divert any cables, pipes and oversails
- cut off supplies
- indicate any which may be or have been removed or abandoned.

Apparatus for non-essential but retained services should be made inert prior to the commencement of demolition; the demolition contractor will require a plan showing the location of cables and pipes, particularly those which are alive or must be safeguarded. An accurate pipes and cables plan should be prepared for the demolition contractors with instructions and details of any special requirements during demolition, as shown in Box 18.3.

Clearance and demolition

Clearance involves the following:

- obtaining possession from tenants and any unauthorised occupiers
- removal of discarded chattels, plant and machinery and deleterious waste
- cutting off of services, such as gas, electricity, gas, telecommunications and water
- erecting any advertisement hoardings
- temporary lettings for income
- access for archaeological investigations (under an agreement)
- fencing or ditching of boundaries
- boarding of apertures in buildings.

Human remains

Situations may arise during demolition and clearance where human remains are discovered, including:

- discovery of remains or a body where the death was recent or of indeterminate time
- chance discovery of remains where archaeological evidence suggests death was in the distant past
- discovery during the clearance of a burial ground
- during formal archaeological work.

In the first two instances above (and in the other instances where the death may appear recent), the police must be informed. They will carry out an investigation and advise the owner and other parties in due course.

Box 18.3 Matters to be considered on the demolition of buildings

Heritage	• check that any listed building (and other) consents have been obtained • physically protect listed buildings and ancient monuments • record for posterity the existing building, etc if necessary
Nature	• comply with any designation of the site and the regulatory regime • physically protect habitats, fauna, etc • as last resort, remove in accord with the authority requirements • observe tree preservation order regime
Archaeological matters	• comply with any designation of the site as one of archaeological importance
Burial grounds	• check local planning authority for evidence of disused burial grounds • if discovered during works, consult with planning and religious bodies • "active" burial grounds are a matter for the religious authorities to deal with
Pipe, cable and oversails	• prepare plan • arrange cut-off of services • protect "active" pipes etc
Infrastructure (on or near site)	• inform relevant authority • comply with standards, as appropriate, eg for railways, ensure site lighting does not affect signalling
Asbestos and other contamination	• comply with regime for removal of asbestos • if known or found, observe "duty of care" and statutory regime for remediation
Planning permission	• check status of proposal regarding planning regime, particularly planning and other consents
Treasure and other finds	• deal with finds in accord with the Treasure Act 1996 and local code for portable finds
Security	• appoint licensed contractor to install security apparatus • ensure personnel have been vetted and in due course are licensed

---

Burial grounds

Where a burial ground is being cleared, the planning and religious authorities should have already given the necessary permissions. The work may proceed in accordance with the formal procedures.

Archaeological remains

Certain areas may be designated as areas of archaeological importance under the Ancient Monuments and Archaeological Areas Act 1979. If so archaeological surveys are usually carried out prior to demolition and clearance, the developer having given a formal notice of the start of development. If an archaeological dig is required, the demolition or development work must accommodate the work for agreed periods, eg up to four months.

Other archaeological evidence, including human remains, is likely to interest the local archaeological body concerned. Normally, human remains are removed from the site for storage and research: to be re-interred in due course.

Insurances

If the original building was destroyed by fire or otherwise, any claim for insurance should be settled. A number of issues should be checked prior to allocating any monies to be received, namely:

- the money may not be available to the owner, but to a tenant or some other person
- the money must be applied to reinstatement
- capital gains tax may arise if delays occur or the reinstatement does not proceed
- the money due may only be available for reinstatement — partly or wholly paying for the redevelopment.

Once a building has been demolished, any outstanding insurance policies should be reviewed for closure.

Landfill tax

Certain developments, eg opencast mineral workings, may have restoration and aftercare conditions attached to the original planning consent. From the 1 October 1999, spoil from the restoration of a site is not charged landfill tax.

Diversion and closure of roads

In Chapter 4 the final section dealt with the ownership of roads and certain aspects of ownership, eg ransom strips.

Private roads

Where the developer owns the property on both sides of a private road and there are no servitudes or rights of way affecting it, closure should be a matter of fencing off the ends and ensuring the security of the boundaries to the frontage property.

Any easements, wayleaves or other third party rights, will need to be extinguished unless the development can go ahead with them in place (see Chapter 4).

Highways

Frontagers normally own the land beneath highways. On a redevelopment, where a developer owns all the property fronting a highway, it is usual for the highway authority to close road under the Highways Act 1980. In some instances it may be appropriate for the road to be diverted.

Contaminated land

A strict regime for dealing with contaminated land is managed and enforced by the Environment Agency and local planning authorities. There are several aspects:

- identification of contamination
- identification of the responsible person
- service of a remediation notice
- remediation works, sufficient to enable development
- monitoring of the site.

Box 18.4 shows who does what in the decontamination of land.

Box 18.4 Roles and activities in decontamination of land

Polluter
• class A person	• if traced, the person who caused or knowingly allowed the contamination and liable for the remediation
• class B person	• if the Class A person cannot be traced, the owner or occupier who is responsible for the remediation of the land
Valuer	• values the contaminated land
Insurer	• may require works and a plan for the future environmental management of the property • may insure the land after remediation
Contamination contractor	• specialist contractor who carries out works to remedy the contamination • may issue warranty or guarantee
Site investigator	• professional who investigates the land and specifies works • issues warranty as to its after-state

---

Identification of contamination

A desk study and site investigation should reveal the presence of any contaminative substances and pollution. The desk study will examine written records, such as:

- public health records of illnesses
- accidents and incidents where spills of substances occurred
- court cases of claims for damages from neighbours or workers on the site.

The site investigation team will undertake initial walk-about inspection looking for the likes of:

- seepage of contaminative substances, discoloured water

- debris from mineral workings, industrial processes or dumping from off-site
- dead or dying plants, trees and animals
- derelict plant or machinery which does or did contain chemicals.

Identification of the person responsible

The appropriate person or persons who caused or knowingly permitted the contamination of land is a class A person under the legislation and is normally liable for the cost of remediation. A class B person may be liable if the class A person cannot be found — the owner or occupier of contaminated land.

Remediation notice

Any classified person may carry out remediation of the contamination of their own volition. However, the authorities may be required to serve a remediation notice.

Briefly, where the classified person or persons are identified the local planning authority will serve a remediation notice requiring decontamination of the land.

Remediation

Some kinds of development will be permitted after a minimum remediation, if any: others will require extensive works to clear the contamination, followed by monitoring. Works of remediation vary with the following:

- the type of contamination
- ground conditions
- given the contamination, the approach to remediation
- the development which will receive planning permission following the works.

Remediation is a complex professional undertaking requiring professionals to investigate the site and prepare a specification for the works which will need to be undertaken to allow a particular development. Sometimes the cost of remediation will not make the development worthwhile unless a subsidy is available.

Capital allowances

The costs incurred in clearing contaminated land are deductible for capital allowances purposes.

Landfill tax

When carrying out remedial works to contaminated land, material removed to landfill is not subject to landfill tax. A Customs and Excise certificate must have been issued by the time the removal is carried out.

Restoration of land

Certain types of land use, such as opencast mining and quarrying, require restoration while the workings are progressing and when they are finished. Aftercare conditions are usually imposed by the local planning authority. Also, where the land is held on a lease or licence by the mineral operator, the landlord or licensor will usually require the land put back to a developable condition or to a condition for agricultural use.

In anticipation of or as part of the restoration, the mined or quarried land may be used for landfill under licence from the relevant authority. Care must be taken to ensure that the waste is suitable for the substrata beneath the area to be landfilled. Obviously, contaminated or active waste must not be deposited if there is any chance that it may contaminate underground waters.

Restrictive covenants

Restrictive covenants are imposed on land in two ways:

- negative covenants — requiring the owner not to do something
- positive covenants — requiring the owner to do something.

This is for the benefit of some other land.

Negative restrictive covenants are used for various purposes which may prevent or restrict what a developer wishes to do, such as:

- limit the development on a large site to one dwelling
- limit the use to retail
- limit the use of a house to use for a family
- prohibit the sale of alcohol or tobacco.

A developer faced with a restrictive covenant, particularly one which is old, may adopt one of the following:

- ignore the covenant and proceed; but then, possibly face a court action
- buy the land which benefits
- take out insurance against possible action, and proceed
- apply to the Lands Tribunal for the modification or discharge of the covenant, perhaps paying compensation.

Procurement of Buildings

19

Aim

To explain roles and practices of obtaining the construction of a building, ie procurement

Objectives

- **to identify the roles and activities in the construction of a building**
- **to briefly describe and compare the organisational arrangements for procurement**
- **to examine the procedures of tendering and agreeing contracts**
- **to consider various perils and how the parties transfer risk by insurance, performance bonds, guarantees and warranties**

Introduction

An account of procurement is essentially about the different ways in which an individual or organisation wanting a building goes about arranging the organisation of the project which will result in its construction and handover, hopefully on time, within the budget and to the required quality.

Roles and activities

The principal participants may differ according to the method of procurement that is adopted by the employer. Box 19.1 is, therefore, somewhat general and the descriptions given below are intended to fill out the gaps.

Project management

A function which must be fulfilled for every project is that of project management, but it may not be possible to identify any one person as the project manager of a particular job (see below). A definition of the term project manager is:

Box 19.1 Participants in procurement — their roles and activities

Employer/Client	• prepares the description or brief of the proposal
Architect or designer	• designs to the brief after developing the contents with the client
Quantity surveyor	• prepares specification and costs proposal • prepares bill of quantities • prepares cost plan
Structural engineer	• prepares structure to support the design
Building services engineer	• prepares specification and plans for the building services
Planning supervisor	• prepares pre-contact health and safety plan for construction, in conjunction with the designer • prepares post-contract building safety plan and manual
Building surveyor	• may prepare design of relatively small projects • commonly, undertakes snagging on behalf of the owner
Main contractor	• constructs the building
Subcontractor(s)	• undertakes work package(s)
Adjudicator	• difference will be referred to adjudicator • decides within 28 days • may seek out facts and law as an expert

---

The leadership role which plans, budgets, co-ordinates, monitors and controls the operational contributions of property professionals and others in a project involving the development of land in accordance with a client's objectives in terms of quality, cost and time (see The Glossary of Property Terms).

Procurement from start to finish

An owner of an estate in land (or land and buildings) may be a developer, in that a new building or an improvement to an existing building may be procured by engaging a contractor or by self-building. In the contractual context the landowner is sometimes called the "client" or "principal" and also "employer". Of course if the client is a leaseholder, all superior estate owners, eg a freeholder, will be held by another landowner but they are not clients. In this chapter the terms, eg employer and client, are used interchangeably with developer, but not for a superior landlord.

New buildings

Having bought the site, raised the finance, cleared the site and, perhaps, obtained outline planning permission for a new development proposal, the developer is well on the way! Procurement is the next phase. Box 19.1 gives an account of the staged approach to completion of a project. It is intended to give an insight to the activities and a feel for the jargon which the parties may face from the start to the finish of the construction phase of a new building or an improvement. There are many approaches to building procurement so the account is schematic and somewhat simplistic. However, as an introduction, it should make the approaches to procurement more readily understandable.

The need for a good group or groups of professionals is essential for reasons of the complexity of construction and the considerable volume of enactments which regulate the way in which construction is undertaken. Some of the legislation is described in greater detail elsewhere in this volume; it includes:

- the Town and Country Planning Act 1990 and associated enactments
- the Building Regulations 1994
- the Construction Design and Management (CDM) Regulations
- the Health and Safety at Work etc Act 1974.

Existing buildings

Chapter 11 outlined the types of contractor and the kind of works which the construction industry undertakes. The above section mentioned the legislation for the procurement of new build; generally, the legislation for works to existing premises is the same but other legislation may need to be considered, including:

- the Party Wall etc. Act 1996 — when the works affect the common wall, the party wall, between properties in separate ownerships
- the Housing Act 1985 — standards for dwellings including multi-occupied premises
- Access to Neighbouring Land Act 1992 — when the works are repairs and involve parts of the property accessible from the adjoining land
- the Fire Precautions Act 1971.

Management — project or construction?

Construction is always managed, but the terms project management and construction management may need to be clarified. Sometimes project management is used to mean construction management. One organisation has distinguished between the two terms as follows:

- project management covers the development process from the idea or conceptual stage, site finding, site appraisal, etc through to commissioning and occupation by tenants (in the company's projects)
- construction management covers procurement by whatever approach from pre-contract preparations through to snagging, remedial works and handover.

In this chapter construction management is taken to mean the latter role and its activities. Of course, the separation of the roles merely reflects the medium (development as a whole or construction) in which the manager's inter-personal and professional knowledge, skills and competencies are applied.

Both kinds of manager work in different, yet familiar contexts. Each will have the following:

- similar inter-personal skills;

Box 19.2 Schematic procurement of a new building or improvement to an existing building

Step 1 Client appoints professional advisors
- architect
- quantity surveyor
- structural engineer

- contracts are drawn up with each
- architect is briefed on requirements

Step 2 Architect prepares sketch drawings and plans

- client considers
- approves, rejects or seeks amendment etc

Step 3 Architect
- prepares drawings and specification

- makes health and safety input
Quantity surveyor
- prepares cost estimate and bill of quantities
- ensure waste management
Structural engineer
- approves plans
Planning supervisor
- ensure health and safety plan for tender etc

- the professionals work in tandem to finalise the building design
- local planning authority office and building control consulted informally on details
- a site waste management plan checklist is prepared

Step 4 Project manager seeks approvals
(PM may be the client or representative
or the architect)

- planning permission
- building consent
- developer seeks any landlord's approval, etc

Step 5 Quantity surveyor or Architect
- finds and select contractor
- obtains client's approval

- prepares tender documents
- conducts tendering process

Step 6 Client and contractor
- enter into building contract
Quantity surveyor

- appropriate contract prepared and agreed
- contractor's insurances, performance bond and other "pre-qualifications' are checked prior to signing
- ensures insurances are sufficient

Step 7 Contractor
- commences work
- appoint subcontractors

- architect inspects work and approves it
- arranges payments by stages
- quantity surveyor costs and approves variations (also client approves)
- building control officer inspects on site at appropriate intervals

Step 8 Building surveyor and contractor
- agree snagging requirements
Planning supervisor
- ensures health and safety file prepared

- agree need for works
- contractor does works
- if satisfactory, building surveyor signs off works

Box 19.2 continued

Step 9   Architect
- signs completion certificate

- ensuring that the work is satisfactory
- ensuring that the remedial works from snagging are done

Step 10   Client
- receives building at handover
- client receives health and safety file

- begins commissioning (see Chapter 20)
- incorporates health and safety points from the file into operational management systems

---

- sets of discrete and common knowledge in inter-personal skills
- common capabilities in management techniques and appraisals
- competency to carry out the work
- different knowledge resources (with some overlap) — based on the development and construction fields).

The common management techniques and appraisals, which will be applied in different contexts include:

- network techniques, such as critical path analysis
- bar charts and Gantt charts
- line-of-balance charts
- budgetary control techniques
- financial accounting and cash flow analysis
- discounted cash flow techniques.

Approaches to procurement

A developer has several approaches to procurement, each with differing characteristics, such as:

- the contractual relationships of the parties
- the configuration of working relationships
- the duration of work from start to finish
- the quality of the completed building
- the cost of the contract.

The approaches to procurement and the characteristics of each are described in Box 19.3.

Tenders

Tendering is a process where contractors are invited to bid for a construction project or other work. The approaches to tendering are:

Box 19.3 Approaches to procurement and their characteristics

Traditional architect
  Positive attributes
- established and known approach
- clear contractual relationships with client for professional and main contractor
- distinct relationships for design liability and for construction liability
- generally, cost known after design phase completed, particularly with competitive tendering
- quality, cost and time controls generally well handled by professional group

  Negative attributes
- start of work must await completion of the design
- duration may be relatively long
- competitive tendering may mean contractor later experiences difficulties at the cost bid level

Construction management
  Positive attributes
- relatively fast overall time
- work packages tendered competitively, so client receives any cost savings

  Negative attributes
- intense level of work for the design team who may lag building work
- cost of management may be relatively high

Design and build
  Positive attributes
- liability of the contractor is clear
- cost should not change significantly
- work can begin before completion of design
- duration may be quicker as a result

  Negative attributes
- final cost not known before work starts
- cost control may be difficult for the client
- quality less well controlled by the client

- open tendering
- selective tendering.

On the one hand, open tendering allows a wide range of contractors to bid for the job but may contain unknown parties. On the other hand, selective tendering is confined to a few known contractors who will be pre-qualified.

Local authorities

Some projects are offered by local authorities on the basis of open competitive tendering. For much of their work, however, they will have prepared lists of contractors who will be invited to bid for particular jobs.

European Union and procurement

While a private employer is generally free to invite or not invite particular contractors, a public employer with a project costing more than a prescribed lower limit must promote it in the Official Journal of the European Union. Registered social landlords are required to comply with the EU procurement regime.

Health and safety

The planning supervisor ensures that all health and safety issues are covered by the designer, contractor and others for incorporation into the tendering process and contract. The contractor ensures on-site awareness and regard for the health and safety plan by all. Similarly, the planning supervisor ensures the preparation of the health and safety file for the client (see Box 19.6 below).

Site waste management plan

Considerable savings are generally achievable by adopting a site waste management plan. It requires careful planning at the design stage and in the pre-contract preparations should ensure that during the progress of construction all parties meet their statutory duty of care in dealing with waste. Although such arrangements are voluntary at present, it is likely that site waste management plans will become a statutory obligation. Box 19.4 gives a conjectural indication of the way the client might expect roles and activities for site waste management plans to be developed by professionals, contractors, local officials and others. (Site waste management plans are in their infancy and some professionals, contractors, subcontractors, planning officers and building control officials have not been fully exposed to the approach.)

Contracts

Schematic contract

A notional full range of contracts would cover every kind of procurement that could be envisaged for construction. In practice, the employer's contract is likely to be one of the following:

- a verbal contract — for the simplest of jobs
- self-prepared contract — perhaps for the straightforward job or for the a major project which does not fit a published standard form of contract
- a standard form of contract, eg a contract developed by the Joint Contact Tribunal (JCT).

Generally, a contract contains two parts, namely:

- an agreement — setting out a brief note of the work or services required and what the parties to the contract will do, eg that the contractor will undertake the work
- the conditions — describing the arrangements for the works and how the parties will act.

This section indicates broadly and schematically what the parties would wish of each other in completing the project on time, to the budget and to the quality needed. Thus, Box 19.5 shows what a

---

Box 19.4 Schematic basis for a site waste management plan (SWMP)

Client	• specify requirement for a SWMP
Architect or designer	• specify or source appropriate recycled materials and components • on refurbishments, ensure reuse of suitable materials by designing in items • provide space for skips for segregated materials • identify waste re-use and reprocessing opportunities on- and off-site
Quantity surveyor	• set targets for cost savings • specify sale and return opportunities with suppliers • specify opportunities for reuse and re-processing on- and off-site
Main contractor	• prepare SWMP • allocate responsibilities to key site workers • brief all workers on the purpose of the SWMP • identify likely sources of waste making • obtain appropriate waste management licences etc
Subcontractor	• see "main contractor" above
Site workers	• allocate roles for waste management • ensure appropriate briefing • monitor progress on site
Planning officers	• hold information on local policies for SWMP and planning
Building control officers	• hold information on local policies and opportunities for supporting SWMPs
Waste management officials	• hold information on appropriate licences for holding and transporting waste • hold information on opportunities for local re-cycling and re-processing materials

---

typical contract or sub-contract might cover. Also, a contract between the main contractor and a sub-contractor will import some provisions or aspects of the contract that the main contractor's made with the employer. Generally, the contract will give the parties and then in articles gives obligations and rights, including references to exhibits and schedules. In effect it is a bundle of documents, including, perhaps, specifications and plans.

Adjudication

Although Chapter 34 deals with dispute resolution, it may be noted that adjudication is imported into construction contracts by the Housing Grants, Construction and Regeneration Act 1996. If a contract does not include adjudication, the Act provides for it.

The impartial adjudicator is required to decide on a matter within 28 days of referral but rather than hear the evidence alone may undertake investigations, ie as an expert, on law as well as fact. The Scheme for Construction Contracts (SI 1998 No 649) contains the *modus operandi* for the adjudicator in the absence of adjudication in the contract.

Box 19.5 Typical conditions or matters included in a building contract or sub-contact (The detail covers both contracts generically.)

Parties	• the employer, main contractor and sub-contactor
Definitions	• defines important words used in the contract
Obligations	• describes what the contractor or sub-contractor will do in terms of service performance, liabilities and indemnities, etc
Compensation	• sets out promise of payment for acceptable work
	• covers rates and prices
	• invoicing procedures
Termination	• gives rights, etc when the main contract is terminated
	• gives circumstances for termination of the sub-contract and procedures
Taxation	• provides for payment of taxes by contractor or sub-contractor respectively (value added tax excluded)
Governing law	• states applicable law, eg English
Disputes	• provides for dispute resolution, appointment of, say, arbitrator
Intellectual property rights (IPR)	• seeks to protect the rights of the parties in their IPR by non-disclosure
	• liability and indemnity afforded to injured party
Confidentiality	• requirement of confidentiality by contractor or sub-contractor (and respective agents, etc)
	• defines information
Notices	• provides for mode of delivery and addresses
Other provisions	• several and joint liability
	• rights of third parties

Standard forms of contract

The employer will usually adopt a standard form of contract for the type of procurement being used, adapting it if necessary. Those of the Joint Contract Tribunal (JCT) are extremely popular, probably being used to the order of nine jobs in 10. The common forms of JCT are:

- JCT 98 Standard Form of Building Contract — with various types of forms for private or local authority clients
- WCD 98 Standard Form of Building Contract with Contractor's Design (1998)
- MPF 03 Major Project Form (2003)
- MC 98 Standard Form of Management Contract (1998)
- MW 98 Agreement for Minor Building Works (1998)
- HG(A) Agreement for Housing Grant Works (2002)
- MTC 98 Standard Form of Measured Term Contract (1998).

Other bodies have prepared standard forms of contract, eg the types of fixed price contracts. Their use is very much rarer than JCT contracts. For instance the guaranteed maximum price contract is probably not used more than 2 times in 100.

Contracts for home owners

The JCT has a number of contracts designed for use by owner occupiers. They include:

- HO/B Building Contract for the Home Owner/Occupier (1998) (where the client deals directly with the builder)
- HO/C Building Contract for the Home Owner/Occupier (who has appointed a consultant) (2001)
- HO/RM Contract for Home Repairs and Maintenance (2002)
- conceivably, JA 90 Jobbing Agreement 1990 ed.

Contractor and subcontractors

Much building work is done by a contractor who is responsible to the employer for the successful completion of the work. Although it is conceivable that the contractor could do all of the work, it is normal building practice for subcontractors to be appointed and employed by the main contractor to do specific work. This enables the main contractor to employ, perhaps, specialist workers and to spread the work over time.

In some instances the employer will be empowered under the contract to nominate sub-contractors to do particular parts of the building. The main contractor employs the nominated sub-contractors and is responsible for ensuring that their work in done satisfactorily. Although the employer should ensure that nominated subcontractors have sufficient insurances. The contractor will also be concerned and check that insurance is in place.

Perils and risk transfer

Construction is a complex set of roles and activities which is often limited to a confined site and a defined working envelope, ie the proposed building or structure. It behoves the employer, contractors and others to jointly and severally to prudently plan the activities to minimise the prospective dangers faced by the workforce and managers and visitors to the site. Box 19.6 sets out a schema for a safe and secure site.

Legislation

A large volume of enactments regulate the working arrangements for construction works on new buildings and existing property, including:

- Construction (Design and Management) Regulations 1994 (CDM) (SI 1994 No. 3140)
- Construction Products Regulations 1991 (SI 1991 No. 1620).

The CDM Regulations are important in that they introduce the planning supervisor who brings planning for safety in the design and site working "interface". Also, covered by CDM is the safety needs of maintenance workers — this is also planned or at the design stage and the planning supervisor is responsible for the manual on maintenance safety.

Box 19.6 Schematic profile for a secure and safe construction site

Security	• access to site is controlled, eg biometric entry only • perimeter boundary is secure and protected • important internal areas are security-contained • alarms and detection etc equipment is installed
Security firm and personnel	• outside security firm registered with the Security Industry Authority (SIA) • in due course, personnel are licensed with the SIA
Emergency	• ensure addresses and telephone numbers of emergency services are in site office and security office • seek official advice on major emergency requirements • have site evacuation plan and any facilities on site in place • brief key staff on requirements • have nominated staff for media liaison and relations
Health and safety, including: • accident • fire • first-aid • reporting • file	• health and safety plan (CDM) is prepared — by principal contributors • induct all workforce in first aid procedures • for five or more employees, comply with the statutory accident book requirements • have HSE reporting forms in the site office and security office, as appropriate • train and update on-site first-aid workers and managers • health and safety file (CDM) is prepared for handover.
Lone working	• lone worker to carry mobile telephone • lone worker to give line manager, security or colleague place, duration and return time • for extended period, lone worker checks in by telephone or, if possible, visited
Protective clothing	• comply with legislation on protective clothing • generally, protective clothing, including steel-capped boots and hard hats • especially, special protective clothing, eg for chain saw workers • ensure clothing is complete and kept in good serviceable condition
Personnel records	• subject to data protection legislation and individual's written consent, have personal medical data available for accident para-medics or hospital doctors
Equipment	• comply with legislation on plant and machinery • protected plant and machinery, eg guards and barriers to moving parts • service plant and machinery in accordance with manufacturer's guidance or to a prepared duration-planned basis
Insurances	• comply with statutory employers' compulsory insurance • post statutory notices on work notice boards • add policy renewal to the caution book • cover for third parties and other all risks type insurances • ensure compliance with insurer's terms and conditions • contactor's and sub-contractors' insurances should be sufficient in each case to cover the whole building

Insurances

The contractor will normally have an all-risks insurance policy as the main cover and other cover to meet perils not included in the all-risks. The cover needs to meet the loss or damage arising from:

* third party incidents, including and in particular those involving children
* employees (employers' compulsory insurance)
* theft of plant and machinery and materials
* fire and storm damage.

In some instances, the contractor's insurer may want the employer's insurer to agree to share responsibility for insurance on a small job in a large building or where the risk is known to be high.

Generally, where the job has a high risk and the potential losses could be extremely high, it might be considered unreasonable for a contractor's insurer to bear the whole risk. For instance, the cover might be shared where a building is worth £200m, the job is priced at £40,000 and the contractor normally carries, say, £40m of cover.

Performance bonds

A performance bond is intended to cover the employer in the event of the contractor not completing the work. The contracting company obtains the bond from its bank or a company which specialises in bonds. The premium is paid by the contractor but adds it to the price for the job. The amount of cover is a percentage of the contract price, say, 25%, and should be sufficient to enable the employer (who holds the bond) to obtain a replacement contractor to complete the partially finished work.

Guarantees and warranties

A guarantee or warranty is cover by a supplier of goods that for a prescribed period an item, eg a building component, plant, machinery or material, is free from defects or will remain in working order, not breaking down. Similarly, workmanship may be covered.

Snagging

The employer or the future occupier should not expect to pay for or accept defects in the building which arise while the work is progressing. Defects occur largely as a result of incorrectly specified or defective materials, poor workmanship or both. Sometimes an accident occurs during construction, eg a burst pipe, which may cause damage not discovered by the contractor. Snagging is the term given to the joint inspection and identification of defects which are noted for remedial work. Typical snags which arise include:

* poorly fitted cupboards or other built-in furniture
* chipped, scratched and stained items
* missing appliances in furnished accommodation
* electrical systems which do not work
* blocked chimneys, pipes, gutters and other conduits.

The potential list is endless and those doing the snagging must know what to expect (at the worst) and have a clear approach to the task.

Snagging is carried out jointly by:

- the employer (owner) or by some knowledgeable and experienced representative
- the contractor or representative
- perhaps, the future owner or leaseholder (where the building has been sold or leased).

The project manager, a professional building surveyor, or an architect could fulfil the role of representative: there are specialist snagging firms and companies offering such services.

A systematic external and internal inspection should be executed. The internal investigation should progress from top to bottom of the building, room by room. Each space should be dealt with in a similarly systematic manner, eg clockwise; the defects being recorded on a proforma. A professional will provide a snagging report detailing the defects and other irregularities. Snagging and any remedial works take place before handover.

Handover

Sometime before handover the owner should begin the preparations to receive the building from the contractor. Concerns which will need attention include:

- entry and occupation before the contractor has finished, ie prior to handover
- insurance
- local taxes
- removal
- connection to services.

The next chapter, Chapter 20, deals with the requirements after handover.

Occupation prior to handover

The future occupier — owner or leaseholder — may need possession before handover. The co-operation of the contractor will be needed and there may be reservations which will need to be addressed, eg security, insurances, expenses arising from the occupation.

Part 6

Property and Facilities Management

Commissioning a Property

Aim

To show what is involved in commissioning a new property

Objectives

- **to consider the contractor's and others' responsibility before and after handover**
- **to show how the new building is brought into property and facilities management**
- **to indicate the range of matters which must be dealt with by the occupier or professionals involved**
- **to describe compulsory purchase involving reinstatement of a home or business in alternative accommodation**
- **to outline system for property management in a workplace or home**

Introduction

After a building has been handed over by the contractor, the owner must arrange for occupation either as an owner occupier or as a landlord by letting to one or more tenants. Each tenant is in effect the owner-occupier of the leasehold property so the roles and activities described in this chapter are broadly applicable: any differences will be explained. The roles and activities for letting a property are dealt with in various chapters (in particular, see Chapters 3, 12 and 15).

Topping out

When a building is completed it is traditional for the workforce to top-out with a celebratory event!

Handover by the contractor

Practical completion

The completion of a building will be confirmed by the architect who when satisfied will issue a practical completion certificate. Handover of the building by the contractor to the employer follows the issue of the practical completion certificate.

Defects following completion

Defects found before and following completion need to be dealt with by the parties involved and a distinction needs to be made between the following kinds of faults:

- snagging faults (see Chapter 19)
- latent or inherent defects
- design faults.

Fundamental defects (latent or inherent) which do not become apparent during construction or the snagging period may be due to one or more of several causes, examples include:

- poor design
- poor materials
- poor construction.

Latent defect insurance should be taken out by the employer before the start of the contract. In the case of new houses or flats the National House Building Council scheme is available for latent defects.

Non-domestic property owner's concern

Whether the owner is a business owner occupier, an investor or a dealer, receiving the building upon completion is followed by a busy time for the owner's representatives who are concerned with property and facilities. Prior to the event preparations will have been put in hand to ensure a smooth transition at handover. For convenience, in this volume the preparations and their execution are covered in three parts, namely:

- property management concerns
- facilities management concerns
- operational management concerns.

In practice the owner's property objective, the size of the organisation and business management policies on property, staffing and outsourcing will determine who will in fact deal with these concerns. The work may be managed by internal management, internal professional staff or outsourced to professional consultants. Similarly, day-to-day operations may be outsourced to one or more contractors or carried out by internal staff.

Investors in property may have property managers and leave the facilities management to lessees. However, the owners of large shopping centres, retail warehouse estates and leisure complexes tend to involve themselves in most aspects of the day-to-day running of the estate.

Property management

Property management operations include the following:

- securing the property prior to occupation
- receiving the owner's health and safety file from the planning supervisor — setting up systems to take account of the file's requirements and recommendations
- relations with organisations in the local community who have a working interest in the property
- establishing the property's records
- planning and establishing control systems for repair, maintenance, security and insurances, together with any other matters covered by the owner's manual
- for an existing non-domestic building (a refurbishment), ensure any asbestos management plan, if required, is in place
- for a multi-property estate, embedding the arrangements for the new building into the estate management systems
- for investment property, liaising with the property marketing staff or estate agents to allow applicants to view the property and carry out inspections with the view to occupation
- arranging the moving in by the owner's staff or by lessees.

Facilities management

Facilities management is concerned with arrangements for the building's repair, maintenance and insurance systems which support occupier's operations in the property, including those for:

- plant and machinery for central heating, electricity, gas, water and other utilities, lifts, external window cleaning, waste disposal and the like
- operational systems for catering, cleaning, information technology and communications, security, waste disposal and the like.

The work will include recruitment, induction, training and development of staff for these systems.

Health and safety

For large enough projects, the planning supervisor is required to prepare a health and safety file covering aspects of the building which the occupier will need in running the building, eg health and safety for operatives involved in maintaining the building. It is given to the employer at the time of handover.

Services — provision

The manager should be assured that all services are up and running and that there has been compliance with the health and safety requirements for their installation and future maintenance. Matters to be covered include:

- services have been checked and approved as being in order

- manuals and plans have been received, recorded and stored appropriately
- building plans and plans of the services tally and are recorded and stored
- maintenance schedules have been prepared
- warranties and insurances have been checked and recorded
- systems for the payment of charges for services, insurance premiums and the like are in place
- where appropriate, back-up services are available and in working order.

Insurances

The manager needs to ensure that appropriate risk assessments have been undertaken to spread risks by insurances or other means. The following insurances and other cover are typical:

- occupier's liability, under the Occupiers Liability Acts 1957 and 1984
- fire, other typical risks and contents
- manufacturer's, installer's or service provider's warranties or insurances for plant and equipment, water tanks and pipes, drainage and the like.

Where leases have been or are going to be granted and the landlord is responsible for insurance, each of the lessees should ensure a note of each of the lessee's interest in the property is placed with the underwriter for noting on the policy.

Sometimes the developer or the owner has a block policy for their estate. If so, the property's acquisition should be noted on the policy.

Taxation

Local taxation

The property manager should ensure that the property is appropriately assessed for local taxation. Normally, a rating specialist's advice would be sought on the assessment by the officers of the Valuation Office Agency.

Business property is assessed for business rates but mixed property (composite property) is treated differently. Where the property is a composite property, only the assessment for the business part is entered in the rating list. Both business rates and council tax will, however, be charged, since the domestic part will be banded for the latter tax (see Chapters 28 and 29).

National taxation — records

The property manager may need to liaise with accounts personnel to ensure appropriate records are kept for national taxes.

Outlays or expenses are normally categorised as revenue costs or capital costs. For taxation, revenue costs are allowed against turnover in calculating annual profits which are then charged to income tax or corporation tax. Capital costs are recorded in capital accounts and kept for two principal purposes, namely:

- where allowable, in calculating any gain or loss in capital gain tax computations on a disposal of the property
- where the expenditure qualifies, in computing an annual writing down allowance or any balancing allowance or charge.

(A fuller treatment of taxation issues is given in Chapters 27 and 28.)

Compulsory purchase

A compulsory purchase usually means that the claimant must find alternative accommodation suitable for the business or the family home. The claimant may build new premises but usually buys or leases the alternative property. The claimant must act reasonably and seek to mitigate the claim.

When a compulsory purchase of business property occurs the claimant is in one of five situations, namely:

- first, the business moves to alternative premises
- second, the business is closed down because the claimant has been unable to find suitable alternative premises
- third, the business closes down because the claimant retires on the grounds of poor health or age (over 60 years)
- fourth, the business closes because the claimant refuses suitable alternative premises
- fifth, the compensation is based on development potential of the property.

In the last situation there should be no claim for disturbance, only the open market value (reflecting development value) together with professional fees and legal costs.

In the other situations the claimant should receive the open market value (in its existing use) together with professional fees and legal costs. Business disturbance will be payable but will differ according to the situation (see below).

Appropriate alternative accommodation

New property which is being commissioned may be the appropriate alternative accommodation which the claimant obtained following the start of a compulsory purchase of the original property.

Property which is compulsorily purchased results in claims for compensation being prepared for the following:

- for leased property, the freeholder as landlord and for each of one or more tenants (each party may have a different professional consultant)
- for freehold property which is owner occupied, the freeholder, or
- for vacant property, the freeholder and any lessees.

Business disturbance

An appropriate claim for compensation for business disturbance when the business moves is likely to cover the following items:

- removal costs
- costs of installing machinery and bedding down the machinery
- double overheads
- temporary loss of profits
- permanent loss of profits
- disturbance costs of removing keyworkers to the new locality
- possibly, the losses from financing the new building on adverse terms, and compared with previous financing terms.

Thus, these items will be payable in the first situation above. In the second and third situations the claimant will receive compensation on the basis of total loss. In the four situations the claimant may have failed to mitigate the loss, the compensation will reflect this and will probably be less than otherwise.

Advanced compensation

It may be possible to obtain up to 90% advance of the acquiring body's estimate of the amount of compensation after the acquiring body has taken possession of the former premises. Interest on outstanding compensation may be claimed from the date of possession to the date of receiving compensation.

Taxation

A liability to taxation may arise on various items of claim for which compensation is agreed and paid, including:

- income tax or corporation tax on temporary loss of profit
- capital gains tax on the consideration received for capital items (rollover may be available to delay payment).

Office or home management system

Sometimes an employee, for instance a personal assistant, is given responsibilities for one or more management aspects of the building. Assuming a non-property person is given (or home owner who has) this responsibility, he or she will need to set up a system which will maximise the potential of the building and grounds throughout the life-cycle of the property. If the procurement is a first-time or one-off, there will be little in place by way of systems or even a place to set up the property office. In some instances, the designated place will, in due course, be the personal assistant's office. The system which he or she might set up will, in principle, be as effective as that in an estate of many properties.

Papers

It is easily conceivable that the new property manager will be handed a small pile of disparate papers, floppy discs, plans, bills, receipts and so on. Box 20.1 gives an approach to sorting out the pile with the view to forming the basis for a system of managing the property and the facilities (for businesses, or even residential owners).

Box 20.1 Schematic approach to setting up a property management system from scratch

Contacts	• set up an address book system for contacts • include names, addresses, role and activities
Data protection	• where personal information is stored, review requirements for the data protection legislation • set up system which complies • issue guidance to all data handling staff • register data holdings, etc
Diary or caution system	• set up diary or caution/warning system • include dates for regular payments, eg rent, rates, insurance premiums, lift maintenance etc • include dates for property and systems maintenance • cross reference to contacts, topic files and plans, etc
Folders and files	• sort the papers into topics for files (and folders, if appropriate) • relate to cost centres, as appropriate
Cost centres	• liaise with accountant or finance officer, if appropriate • prepare periodic budgets • establish cost centres • prepare formats for accounts and payments systems, together with taxation
Emergencies and risk management	• analyse risks to business (or home) • establish hard risk containment — changes to the building and facilities (involving works) • establish emergency procedures for staff and visitors etc • update contact list with outside services, etc, eg emergency and rescue, police
Evacuation	• liaise with fire service personnel, police crime prevention officer and other local advisors • establish procedures for evacuation and other emergency situations • promote the procedures to staff • train, etc key staff, if necessary • programme fire and other drills (update diary) • hold trial evacuations, monitor and evaluate
Insurances	• ensure compulsory insurances are in place at all times (note diary) • given risk profile, establish current insurance cover • review insurance needs and seek advice on best cover available • check compliance with all the insurer's requirements • ensure compulsory notices are in place and up to date
Property and facilities programmes	• establish programmes for repair, maintenance, etc to property and building services • establish facilities cleaning maintenance, etc programmes • update diary
Storage	• prepare storage records • prepare stored items renewals plan, eg hazardous materials

The box is not exhaustive but is indicative of a simple system for good stewardship of the property, facilities, contents and, particularly, for the safety of one's colleagues or family and self.

The next chapter examines this topic in much greater detail.

Caring for Property

21

Aim

To explain the need for and execution of the stewardship of property

Objectives

- **to describe the roles and activities of estate and property management and facilities management**
- **to show how different businesses require different kinds of management**
- **to specify the management for different states of a building's life**

Introduction

Caring for property or the stewardship of property is multi-functional and covers two professional fields — property management and facilities management. The latter has been developing quickly into a fully fledged professional and technical field in its own right for nearly 20 years. A simple division of the two functions is attempted here but there are many nuances of approach which do not fit a simple approach.

In its purest operation property management covers the landlord who manages, or has representatives to manage, an estate of properties which are let on full repairing and insuring leases. The tenants are responsible for the day-to-day property management but are also responsible for the properties' operational systems to support the occupation as well as the business systems themselves. For instance, facilities management services the following include, among others:

- repairs, maintenance and insurance of the buildings
- establishing and maintaining security, catering, health, safety, nursery and welfare systems
- establishing and maintaining information and communication systems, mail room services and document storage and retrieval
- space planning and the management of churn.

Of these, only the first is normally regarded as property management unless the landlord is concerned with multi-let property. Here the management tends to cover many facilities management services although it will still be called property management by some professionals. In other words, there is no clear dividing line and job titles in different organisations may be different for similar or the same roles.

Roles and activities

At its simplest level managing the care of an estate has two main functions, namely:

* property or estate management
* facilities management.

The common perspective of the roles of the property or estate manager and the facilities manager is given in Boxes 21.1 and 21.2. However, both roles may be perceived in a broader conceptual framework as shown in Box 21.3.

Box 21.1 Property manager's role and activities

Relationship between the owner and the occupier(s)	management: • grants of new leases and licences • renewals of leases • repairs and maintenance • provision and maintenance of building services • insurances strategy and implementation • works arising out of damage or loss covered by insurance • minor improvements by landlord or tenant right to manage • take into account any statutory requirements when tenants of a block of flats exercises the right to manage
Estate accounts and finances	management and liaison with estate accountant: • budgets for the individual properties and the entire estate • collection of rents and arrears • payments and receipts repairs and maintenance • service charges, sinking funds and renewals funds • dilapidations receipts from tenants • insurance claims
Taxation	management and liaison with estate accountant: • tax planning to obtain exemptions and reliefs • records for revenue and capital receipts and expenditures
Building and estate design	management and liaison with designers, landscapers, etc: • contribute to the design and evaluation processes • contribute to tax planning at feasibility stage • establish and maintain records for capital allowances
Estate development or redevelopment, etc	management and liaison with participants: • prior to development

---

Box 21.2 Facilities manager's role and activities

Relationship with the owner and business	management — liaise with suppliers contractors and professionals • establish the space requirements for the business • procure space or dispose of surplus • procure furniture, equipment and services for the space
Accounts and finance	liaise with finance officer • set up and maintain records for services • categorise expenditure to cost centres for analysis and tax planning
Taxation	liaise with finance officer • establish tax plan • run operations within the plan
Facility development or redevelopment	liaise with designers • others in formulation a specification for space and usage • monitor and control progress
Business support services Staff support services Building support facilities	liaise with management and suppliers • procure information and communication technology facilities • procure catering, recreational, health and safety, etc facilities • procure security, gardening, porterage and the like facilities

Basis for the configuration of management

How an estate configures the activities (and hence the roles) will depend in a general way on the following:

- the owner's status and objectives in owning property
- the size and geographical distribution of the estate
- the pattern of tenures and occupations of the buildings
- the condition of the buildings and structures
- the financial resources available to manage the estate.

In considering property management and facilities management below it would be helpful to appreciate that the separately presented management may in fact be combined in the management of a particular estate. For instance, the competent owner-occupier of a small building — a house, shop or workshop — may carry out all management activities, even carrying out repairs and maintenance.

Need to manage property and facilities

The fabric of buildings and structures, whether owner-occupied or tenanted, requires repair and maintenance. Cleaning, properly functioning lifts and other mechanical services, porterage, reception, security, telecommunications, utility services (electricity, gas and water) and catering are a few of the many building services, operational support services and staff support services required by users of

---

Box 21.3 Checklist for property and facilities management

Tenure
- check the terms and condition of any lease or tenancy
- record responsibilities
- put actions in the diary or cautions file

Repair and maintenance
- building structure and fabric
- common parts
- plant and machinery
- interior of accommodation
- check the lease for responsibility for any or all of these items
- if the landlord's responsibility, is any service and renewals fund provided for in the lease
- if the tenant's(s'), what access rights does the landlord have to monitor performance
- check the records for future inspections and payments under any guarantees, warranties and the like

Insurances
- fire and other all risks
- lifts, boilers, etc
- check the policies to ensure that they cover the risks the freeholder and tenant(s) do not wish to self-insure
- check that all terms and conditions have been complied with, eg alarms, means of escape etc are statute compliant or to the insurer's specification, eg levers for keys
- check that the insurance certificate is posted for employees, etc

Energy management
- consider sustainable principles for energy (see Box 14.1)

Water management
- consider sustainable principles for water (see Box 14.1)

Outside areas
- garden and grounds
- car parking

- check arrangements for garden design, maintenance
- garden water management — consider xeriscaping
- check responsibility for day to day management
- check any need for working parking levy management

Neighbours and others
- check any rights of neighbours and others which may affect the use and occupation of the property, eg a restrictive covenant

---

property in most sectors. Checklist Box 21.3 reviews a range of property and facilities management functions, but does not consider functions for operational support facilities or staff support facilities in any detail.

Outsourcing

Outsourcing of property or facilities management has grown in recent years, with contractors offering a full range of services or specialist contractors offering one or, at least, a limited number of services. Typical service segments in this field of contracting include:

- building repair and maintenance
- catering
- cleaning
- gardening services, eg design and maintenance
- interior landscaping and plant care
- property lettings management
- security.

Outsourcing companies make offerings of repairs and maintenance to buildings and building services. Others offer a much more limited range. There is a fairly vigorous consolidation — mergers and acquisitions — in many of the sub-sectors in this field.

Tenanted land and buildings

Tenant's burden

The lease should specify who will be responsible for undertaking and for bearing the costs of repairs, maintenance, insurances and operating services. Normally, the landlord will seek to shift to the tenant (or tenants) the burden of undertaking and hence paying for property outgoings and services. To the extent that the landlord is responsible for such items, he or she will want the tenants to carry the cost.

Property let to one tenant

Where a building is let to one tenant and the landlord is not responsible for the management of the building on a day-to-day basis, ie the lease is full repairing and insuring, the landlord's concern is the protection of the investment, essentially a supervisory role. The landlord's management activities in this instance are shown in Box 21.1.

Multi-tenanted property

Where a building is let to several tenants, the landlord will normally have a wider range of responsibilities. However, even here the lease may provide for a tenants' organisation to act on their behalf. Box 21.4 shows several different ways in which the care of the property is conducted in multi-tenanted property.

Recovery of landlord's costs

The lease will normally provide for the landlord to recover the cost of providing any services to the tenants. Box 21.4 gives brief details of three ways in which this may be done.

Vacant property

Vacant property awaiting demolition is dealt with in Chapter 18. In this section, vacant property which is not to be demolished is dealt with from a life cycle perspective. It touches upon such matters as

---

Box 21.4 Recovery or shifting the costs of caring for tenanted property

Inclusive rent	The rent is inclusive of all the landlord's costs of repair, management, insurance and services. Landlord is in control but rent may be insufficient in a particular year.	The expenditure is normally allowed against gross rent from the property for income taxation. Any loss may be allowable against rents from other property let by the landlord.
Service charge	The estimated service charge is paid in addition to the rent. It is collected with the rent and adjusted at the end of each year to reflect actual expenditure incurred by the landlord.	The service charge may include an allowance for the landlord's management of the services. For income taxation, the landlord will normally be liable on any "profit".
Renewal fund (or accumulation fund)	Major items may not need repair or renewal for several years. The lease may provide for an accumulation of funds year by year.	The annual payments may be liable to income taxation in the hands of the landlord. Planning may be needed to structure the arrangements to properly avoid any taxation.
Tenants' association	The intermediary association undertakes works and so on, collecting monies from the tenants for payments to the landlord, contractors and others.	This approach needs to be properly structured but should be tax efficient, being a mutual body on behalf of the tenants. The payments made by the tenants will be allowed for their income taxation assessments.

---

ownership and possession during the life of a property, either as land and buildings or a site without buildings.

Vacant property for sale

Four main periods are considered here, namely:

1. property on the market prior to contract
2. property contracted for sale
3. property after completion
4. property where possession has been taken prior to contract (or prior to completion).

Box 21.5 covers the various periods during a sale (both freely on the open market and on compulsory purchase. It gives the points which need to be considered on such matters as insurance, security, maintenance and, in the event of a compulsory purchase, any entitlement to advance compensation and interest.

Box 21.5 Management of vacant property in periods prior to its sale completion

Property on the market prior to contract	• seller is in possession and responsible for insurance, repair and maintenance • seller responsible for security • estate agent may have keys and gives access to applicants
Possession taken under a vesting declaration (a compulsory purchase procedure)	In a compulsory purchase, the buyer (acquiring authority) in effect "completes" and takes possession under a vesting declaration • buyer responsible for security • seller's responsibility for the property ceases • seller may claim advance compensation and/or interest
Property contracted for sale	• seller is in possession • seller is responsible for insurance, buyer's interest noted on the policy • seller repairs and maintains • seller responsible for security • seller or estate agent may give access to buyer • seller must not remove fixtures, fittings or plants etc unless the contract provides for their removal
Possession of the property taken prior to completion	In a compulsory purchase buyer (acquiring authority) takes possession, non-vesting declaration cases: • seller may claim advance compensation and /or interest • prior to possession, seller may only remove what has been agreed • buyer responsible for security • seller's responsibility for repair, maintenance and insurance ceases
After completion (and after vesting — see above)	Buyer (as the owner) is in possession • buyer fully responsible for the property

---

Special situations

From time to time a property becomes vacant as a result of some event. Such occasions may call for special management strategies to be adopted. For instance, a hospital building or a clinic may be closed for specialist cleaning as a result of contamination by infectious disease or a break-down of an essential building service.

The manager should, ideally, have emergency or contingency plans prepared to deal with such situations.

Insuring against Perils

Aim

To explain how insurance works for property

Objectives

- **to identify the roles and activities in property insurance**
- **to show how the risks may be mitigated**
- **to establish the principles of insurance**
- **to describe the principal types of cover**
- **to outline the approach to valuation of property for insurance**
- **to outline a basis for a claim**

Introduction

The main peril which buildings and structures face is fire but there are many more. In this chapter an attempt has been made to put the various insurances in their contexts. Risk management is a starting point but the chapter considers the principal perils and how to reduce the risk of an event which will result in damage or destruction. Insurance has been dealt with in many other chapters — they will be referred to as appropriate.

Roles and activities

Those in the insurance industry who are involved in property related insurance are briefly described in Box 22.1.

Nature of insurance

The person wanting protection, the insured, obtains cover from the insurer, whereby the insurer will indemnify the insured against loss due to a specified peril (or perils). The building owner or occupier

---

Box 22.1 Roles and activities in the property insurance industry

Insured	Seeks insurance for indemnity in the event of loss due specified perils resulting in: • injury or death to persons • destruction or damage to tangible or intangible property
Insurer	Offers cover against loss due to specified perils
Broker	Advises insured on the types and availability of cover
Loss adjuster	Advises insurer on a claim and will seek to settle it with the insured
Loss assessor	Advises the insured.
Valuer quantity surveyor or building surveyor	Advises the insured on • bases of value of property for indemnity or reinstatement • figures for a claim for reinstatement or indemnity
Insurance inspector	Advises insurer: • on prospective insured • assesses risk of the property and use • on recommended works to property • on the insurance rating
Accountant	Advises on : • financial consequences of disruption to the business • taxation consequences of receiving indemnification compensation and its application
Building surveyor development surveyor or architect	Advises on • the reinstatement of a damaged building • planning controls for the new building • the redevelopment of the site after total loss
Property manager or facilities manager	Advises on the property consequences following an event, such as: • suitability of alternative accommodation for occupants of the building • offering accommodation to any tenants
Risk manager	• assesses risks • recommends works, training and other means of reducing risk
Insurance Ombudsman	Hears complaints where insured used the insurer's complaints procedure without satisfaction • voluntary, not all insurers belong • awards binding on insurers up to specified limit

---

- buildings and structures
- a building's plant and machinery or fixtures and fittings
- furniture and equipment used in the occupier's business
- work in progress and stock
- business intelligence and records.

Finally, but by no means least, a peril may result in injury to or cause the death of employees or visitors.

Risk management

The individual or organisation seeking insurance should prepare the business, property and any personnel in such a way that should an event occur which is likely to cause damage or destruction, the damage is avoided or at least mitigated. Box 22.2 shows the ways in which a prudent insurer will seek to manage risks and, hence, reduce premiums. Although the box relates to businesses, the owner of a dwelling will find many of the points relevant to home ownership.

Principles of insurance

Case law on disputes about insurance has established a number of principles of insurance which must be observed to ensure that a policy is valid. Generally, a challenge by an insurer to a claim by the insured will be successful if the insurer can show that the insured has breached one or more of the principles set out in Box 22.3.

Principles — case law

The principles are derived from insurance practice and case law. Some of the cases are given in Box 22.4 to illustrate the approach adopted by the courts.

Acquisition of property

Gazumping

In a competitive market it is not unusual for a prospective buyer to have an offer for a property gazumped even though the seller had accepted (subject to contract). Gazumping occurs when another prospective buyer offers a higher price and this price is accepted. It may be possible for a prospective buyer to obtain prior insurance against losses, ie abortive agent's fees and legal expenses, which arise when he or she is gazumped.

Contract to completion

A property buyer's and seller's insurance requirements during conveyancing need to be considered in some detail. Briefly, the seller is responsible until the completion but the buyer should be noted on the seller's policy. A similar situation may arise when an owner agrees to lease a building. Similarly, a dormant restrictive covenant may not always remain dormant — the buyer will want to ensure that any loss arising when the owner of the restrictive covenant successfully prevents development, will be covered by insurance.

Blight

New buyers (and existing owners, perhaps) may wish to obtain insurance against blight. Blight is loss which arises when a public body or other organisation proposes a project which leads to a reduction in the value of property or some other loss. The current issue is the prospect of the construction nearby

---

Box 22.2 Checklist for managing risk for insurance purposes

Compulsory insurances	• comply with statutory insurance requirements, eg employee cover, traffic third party insurance • display up to date post workplace insurance certificate on notice board
Induction, training and development of staff	• induct all staff in the use of alarms, the means of escape and other emergency procedures • train all staff in the use of safety and escape equipment • develop "safety" culture and attitudes, eg forethought
Buildings, plant and machinery and grounds	• design and build to health and safety standards • comply with any health and safety working standards for the business • comply with access requirements for persons with disabilities • ensure appropriate storage for hazardous substances • ensure that escape routes are clearly marked • properly maintain buildings
Vehicles	• ensure compulsory insurance cover • ensure repair and maintenance is prompt and fulfils statutory requirements as a minimum • train and develop drivers
Managerial systems	• have directors, managers and others with clear roles and policies for health and safety • establish management procedures for action in emergencies • ensure accident book and accident reporting runs properly
Information and communication technology systems	• comply with 1984 and 1998 Data Protection Acts • display certificate for compulsory staff insurance • have clear and transparent policies for personal use of the ICT equipment
Health and safety	• give training in handling and emergencies involving hazardous substances • check insurance policy on smoking and have a clear policy on it, eg eliminate possibility of non-smokers receiving smoke from smokers, have set-aside rooms or spaces for smokers
Security	• ensure any contractors are licensed and recruit licensed staff
Property transactions	• when buying property ensure that the property is insured and the buyer's name is noted on the policy (see below) • when leasing ensure the tenant's position is safeguarded (see below)
Procurement of buildings	• ensure that the professional consultants and the contractor is covered • even small jobs may require substantial cover (but the employer's insurer may agree to take the burden in a given case) • ensure design covers health, safety and welfare
Environmental land management	• ensure that the occupiers of property, whether owner or leaseholder comply with the insurance policy and have environmental management systems in place
Easements or restriction	• if an easement might be breached, ensure insurance cover is sought (see Chapter 2)
Principles of insurance	• observe the principles of insurance — so as not to invalidate the policy

Box 22.3 Points to be aware of in insurance (based on the principles of insurance)

Good faith	• the insured must supply all the information that the insurer needs to determine the policy, if any
Insurable interest	• the insured must have an interest in the property which can be insured
Indemnity	• the insurer undertakes to put the insured in the same position after the event
Reinstatement	• reinstatement of a damaged or destroyed building usually satisfies the indemnity principle • alternative approaches sometimes apply
Subrogation	• the insurer allows the insurer to take action against a third party in joint names
Contribution	• cover with more than one insurer at the same time will result in the insurers sharing any claim
Average	• if under-insured, the insurer reserves the right to pay a proportion of the loss
Betterment	• on reinstatement, any increase in the value of the property will be deducted from the insurance monies
No claims	• where the insurance does not make a claim in a specified period, a deduction from the premium payable is allowed
Self-insurance	• self-insurance is the result of in-action, inadequate cover or but more appropriately careful appraisal of need

---

Box 22.4 Cases on the principles of insurance

Indemnity not reinstatement	*Leppard* v *Excess Insurance Co Ltd* (1981) Policy provided for reinstatement which was claimed after a fire. Cost of reinstatement (£8,694) more than twice the loss of market value, ie MV (£4,500) - Site Value (£1500) = £3,000 *Award:* = £3,000
Reinstatement not indemnity	*Reynolds* v *Pheonix Assurance Co Ltd* (1978) Judge found evidence difficult to reconcile. Initial cover £18,000 but increased to £550,000 for building (add other items). Property destroyed. Valuations by Plaintiff's two valuers £240,000 and £180,000. Defendant's valuer £25,000 (Site £20,000). Also, modern equivalent reinstatement was £55,000. Reinstatement: £315,671 without recycled materials £246,883 with secondhand materials etc. *Held*: £246,883 + VAT + professional fees (12.5 %), without deduction for betterment
Betterment	*Michael* v *Ensoncraft Ltd* (1990) Investment property worth £140,500. Regulated tenants left after fire caused by defendant contractor. Plaintiff owner offered £203,000 by potential buyer. Owner claimed £200,000 + nuisance etc against contractor *Award:* £136 (temporary repairs) - owner gained from negligence of defendant.

of a wind farm (and concomitant transmission lines) or a mast for mobile telephone transmissions. Of course, the insurance must be taken up before there is even a hint of any project!

Works to a building and cover

The building or construction contract will or should ensure that the contractor has insurance for the building according to the normal contractor's all risks policy. Where the employer requires other kinds of cover these should be negotiated at the outset.

Insuring buildings

Normal buildings

All buildings are rated by the insurer according such matters as:

- the kind of construction
- the use to which the property is put
- the contents, eg in terms of physical protection afforded by the building
- the level of physical security and operational systems for security.

Also, for a building in multiple uses which are relatively benign, the insured may find that on a reletting of one unit, the change of use may result in a substantial increase in the premium rating for the whole building, this may happen where the new use is high risk.

Obsolete buildings

With the agreement of the insurer an obsolete building may be valued as a modern substitute building, particularly when on the total loss of the property the local planning authority would not allow a rebuilding of it.

Historic buildings

Historic buildings can cause problems for the insured. For instance, on a total loss of the building, the insured may not wish to reinstate the building — not wanting a pastiche. Also, historic buildings are relatively very expensive to reinstate. Where there is a partial loss, there comes a point when a building's cost of repair becomes too expensive to insure. Here the "first loss" becomes available (see below).

Valuation for insurance

Four broad approaches are recognised for insurance valuations (only one of which is a true valuation approach rather than a cost approach), namely:

1. reinstatement (a cost approach)
2. simple substitute building (a cost approach)
3. market value (a valuation approach)
4. first loss approach (a cost approach).

Reinstatement approach

The approach is simply an assessment of the total cost of replacing the insured building. Box 22.5 shows many of the issues likely to arise in a reinstatement of a destroyed or damaged building. If in the event, it is appropriate to use a typical reinstatement valuation, the works indicated by an asterisk would normally be taken into account (the valuation will include other matters, eg the cost of building works).

Box 22.5 Points to be taken into account in considering reinstatement

Valuation	• type	• reinstatement may not be appropriate
Building	• listed	• listed building consent may be needed for demolition and works
Location	• conservation area	• conservation area consent may be needed for demolition and works
Demolition*	• preparations	• plan showing pipes and cables will be needed • arrange for services to be cut off • contractor's insurances in place • if needed obtain planning permission
Making safe and site clearance*	• preparations	• see Chapter 18
Professional services*	• fees	• roughly 12.5% to 15% of the cost of works
Planning and building control*	• fees	• assess likely cost
Property adjoining*	• protection	• support • weather proofing • fencing • party wall negotiation
Debris*	• damage caused	• make good damage to sheds, fences and so on • part of clearance work
Access and storage*	• restricted site	• contractor needs to find off-site facilities
Features of old building*	• special	• arrange to restore and reinstate • architectural salvage, ie re-use or sale of items
Fixtures and fittings*	• ownership	• for let building, landlord's property perhaps — reinstate

---

Simple substitute

Where the building is obsolete, eg a 19th century mill, and would not or could not be replaced, the approach is to assume a substitute modern building would replace the original. The assumed building's specification is modern, meeting current standards, and it is likely to be smaller.

Market value

Where the insured building is worth considerably less than its site, insurance based on a reinstatement approach would result in the insured being in a better financial position after receiving the insurance monies.

First loss

First loss approach is commonly used where the total reinstatement of the insured building would probably be regarded as inappropriate. Two examples illustrate this concern:

1. cost of reinstatement of a very large building would result in premiums which the insured may consider excessive
2. total destruction of a historic building would result in a pastiche or "fake".

However, such a building might be damaged or partially destroyed and the insured would wish to reinstate. In these circumstances, the owner decides on the notional level of damage or partial destruction at which reinstatement would be carried out. The result may be that up to say 40% of the building would be replaced by the insured. Insurance is then based on this level of reinstatement.

Inflation and valuation

The date of valuation for insurance will be shortly before the commencement of cover, ie the basic reinstatement cost will be that at the commencement of cover. However, a valid claim may arise on the last day of the period of cover, ie a year later — there should be an allowance for that year's expected inflation. Similarly, construction takes time — building may take, say, three years to procure — so a further inflation allowance should be made at a forecasted rate of inflation. The appraiser will need to allow for the, possibly, variable rates of inflation. The case of *Gleniffer Finance Corporation* v *Bamar Wood Products Ltd* (1978) allowed a period of 2.5 years for inflation.

Contents and other cover

In addition to the building, the business occupier, as the insured, will want cover for such items as:

- furniture and ICT equipment
- building services and plant and machinery (if not covered by the landlord)
- cash and intangible valuables
- raw materials, work in progress and stock.

In the case of plant and machinery, where they are not part of the building, there may be insurances with the supplier, eg for boilers or lifts. Obviously, the large windows are part of the building. It is common, however, for the shopkeepers to insure them separately as they are particularly vulnerable and, if covered generally, may affect the no-claims record.

Disaster recovery cover

A disaster will not only result in the normal losses indicated above, but may imperil the survival of the business. Many organisations therefore seek substantial kinds of consequential loss cover. Insurers are willing to go along with this cover but may impose strong conditions. For instance, they will want secondary "dormant" premises for an immediate move. If necessary, the premises will contain or have access to business data systems and hence operational ICT. (This may also be a requirement of the Financial Services Authority for some financial institutions.)

Personnel and others

A duty of care, in effect arises under a number of statutes and this duty is discharged in part by owners, occupiers and employers ensuring good management systems, the proper care of buildings and equipment and so on. Also in part they will use insurance. Of course, the employers' insurance for the employees is compulsory but there may be a need for:

- keyman insurance
- health care insurance
- third party (to cover customers, visitors and contractor's personnel)
- public liability cover.

Some insurers supplying health care insurance also offer back-up medical and restorative facilities to personnel who become injured at work. Finally, national health insurance contributions are probably seen by some as a tax rather than insurance.

Guarantees, warranties and performance bonds

Performance bonds

Employers of building contractors usually require a performance bond against any loss which may arise if the contractor cannot complete the job.

Latent defects insurance

The buyer of a new home will usually find that there is a guarantee in place for, say, 10 years against structural defects. The National House Building Council runs a national scheme of insurance to which many builders belong, namely "Buildmark".

Installations

If a guarantee, surety or warranty is in place, it may not be necessary to insure an asset against a peril. Guarantees and warranties are common for some kinds of service equipment which are installed in buildings, eg lifts, boilers. If something goes wrong within a given period, they are repaired or replaced. Similarly, a timber floor affected by dry rot may have been replaced. Here, a warranty may have been supplied by the installer, but for a limited period, say 10 years.

Also, service agreements may contain a charge for the annual service of, say, a boiler but also an insurance element for replacement of parts. The home owner may now obtain service agreements for most utilities, ie gas, water, plumbing, drains, ICT, electricity and kitchen appliances.

Making a claim

Following loss of, or damage to a building, the insured needs to prepare and make a claim to the insurer. Box 22.6 provides a schematic approach.

Box 22.6 Schematic approach to making an insurance claim

Step 1a	Insured	• notifies insurer of loss or damage • supplies insurer with immediate documentation • prepares file and accounts for claim record and submission • appoints loss assessor to advise • appoints building surveyor and other professional to advise on building works and other losses • notifies other parties to the policy (if any)
Step 1b	Insured	• prepares setting up or alternative arrangements • arranges removal and storage etc
Step 2a	Loss assessor	• receives instructions • reads policy and relevant documentation • confirms claim can be made
Step 2b	Building surveyor and others	• makes qualitative lists of loss and damage and tentative estimates of losses • prepares reinstatement cost estimates (or other insurance appraisals)
Step 3	Insurer	• sends claim documentation • arranges for loss adjuster to take case
Step 4	Representatives	• seek to agree arrangements and basis for claim • agree heads of claim
Step 5	Insurer and professionals	• reinstate • prepare and submit claim
Step 6	Insurer	• agrees claim and pays (or parties negotiate to agree or resolve any dispute)

Part 7

Property and Finance

Financial Performance and Measurement

Aim

To examine the way in which performance may be measured in the property industry

Objectives

- **to review the principal functions of property ownership**
- **to highlight the features of an estate or property business plan**
- **to examine in principle the various techniques of measuring performance**

Introduction

Every prudent owner of an estate, which is taken here to include a single property, will evaluate the estate from time to time. There are numerous ways of assessing the quality of an existing estate or its potential to be changed successfully. This chapter relates the actual or potential owner's need for insights into the manner of making an assessment. Essentially the owner needs to know what to look for and to understand the outcomes of a professional advisor's work. However, before the professional advisor starts the evaluation, the client may or should expect some detailed self-questioning of the context for any proposals. This will cover the owner's objective in owning the property and the way in the functions of ownership will be addressed (see Chapter 6).

Addressing the estate is a question of the owner's motivation in, and ability to direct and manage the future of the estate, perhaps in terms of the lifecycle of every single building. This will apply whether the owner is an individual or a company or some other body. The owner's performance must be measurable.

Another aspect of the property market is the takeover of a company. The target company may be a property company, a contractor or another company, eg a manufacturer, which operates from a portfolio (estate) of properties. Those targeting the company need to evaluate organisational performance and undertake "due diligence". This chapter endeavours, therefore, to outline the approaches to measuring the ways in which companies and other bodies expect to perform against their actual performance.

None of the above is entirely quantitative: qualitative assessments may need to be made and approaches are described.

Roles and activities

Box 23.1 provides an insight to the roles and activities in measuring financial, business or investment performance. Some of the roles overlap in some activities.

Box 23.1 Roles and activities in measuring property and business performance

Owner/developer	• clarify objective for owning property (see Chapter 6)
Valuer/appraiser	• advises on values, development appraisals, funding situations
Accountant	• prepares accounts and budgets • analyses company's and other bodies' accounts • advises on financial and taxation implications of proposals • carries out due diligence studies of companies and other bodies
Financial analyst	• analyses company's accounts, trading potential in the business context • reports on findings
Financial publisher	• publish monthly reports on companies on the various stock exchanges, eg Really Essential Financial Statistics (REFS)
Technical analyst	• uses trends in share price movement and other information to appraise potential company share investments
Auditor	• reports on a company's preparation and presentation of its annual report
Quantity surveyor	• advises on contractor's pre-qualification for a construction contract • prepares estimates of the cost of construction form the designer's drawings • prepares cost plan, budget for construction and life-cycle costs • advises on development's construction progress
Funder	• evaluates the proposed purchase or development from a funding perspective
Stockbroker	• prepares prospectus for listing purposes (complying with the Purple Book)
Sponsor	• advises on investments in shares
Credit rating agency	• publishes a code of investment risk AAA etc, BBB, etc • investigates a company and makes an assessment • advises of company's investment potential in terms of the risk to the funds invested

---

Functions of ownership

In Chapter 6 the purpose of owning property was described in terms of business occupation, residential occupation or one of the other objectives. Every owner acquires a property at some point in its lifecycle. During a long period of ownership the owner, perhaps a family, a company or a trust, may see the property through a complete lifecycle. Evidence of good stewardship of the holding will be derived from an appraisal of the functions of ownership. For any estate these may be listed in brief as:

- acquisition
- development (major work)
- property management
- repair and maintenance
- improvement (minor works)
- disposals
- financing
- resourcing.

A review of functions for a property, group of adjacent properties or geographically dispersed properties will give insights as to the future of the estate.

Acquisitions

Whether it is a site or a building, a specification for property purchases will usually be in the mind of the prospective buyer, but preferably written. The thinking will be akin, if not the same as that required to determine a marketing mix for end users of a property and others (see Chapter 5). Each prospective property will be considered measured qualitatively and, if possible, quantitatively against a predetermined want list.

Sometimes a purchase will be opportunistic. Unexpectedly, a property abutting an estate property comes on the market; it must be snapped up without too much analysis. Preferably, the estate property's property business plan will contain a section on adjoining properties and the *modus operandi* for ideas about dealing with them. The essence of such opportunities is the synergetic effects which may arise, such as:

- marriage value
- increased density of development for redevelopment
- improved or additional access to the estate property
- merger of interests and possible extinguishment of servitudes burdening the estate property.

The objectives of ownership will determine many of the items on the list, eg a dealer intending to sell after improvements, will assess, say, the profit, the break-even, the internal rate of return and so on. Actual calculations will be made by someone with the perspective of this chapter but often intuition or a hunch will work for the entrepreneur — a different approach.

Major developments

The marketing approach and most of the development stages (given in Chapter 15) will provide the framework for the measures required for a successful development. An investor, for example, will use or attempt to assess such measures as fitness of each building function for a particular market segment (the end users), prospective return on capital and internal rate of return. For an owner occupier, the internal rate of return and fitness for the business as measured by the specification or business model are possible measures. Here, however, the specification is likely to be more detailed than an investor might require.

Property — business or estate plan

At any time every property will have on-going property management functions being carried out, such as rent collection, rent reviews, maintenance and emergency repairs, insurance and re-letting. For the owner-occupier or the occupying lessee, business support functions will be operating, eg security, gardening, reception, post room services, heating and lighting. The owner may or may not provide the lessee with any of these.

However, although these functions are measurable (see below), the measures would not reveal the full picture for the property's future: a more active management plan is needed, and that needs to be appraised and written up as a regularly reviewed property business plan.

Property — costs service

Benchmarking

Measuring performance of something against a peer — benchmarking — is common practice throughout the property industry. So measuring the estimated cost or actual cost of, for example, a service or building is very common. A number of cost information services are available to the developer, landlord or occupier. For instance the RICS offers two comprehensive series:

- the building cost information service (BCIS)
- the building maintenance cost information service (BMCIS).

The former comprehensively analyses the cost of recently constructed buildings by floor areas and, among other items, by components. In effect, anyone planning the design, costing and construction of a building has ready-made benchmark system in place. The other looks at building maintenance costs.

Corporate performance

Many have a stake in the future of a company or other body, including shareholders, taxation authorities, staff and directors, lenders and suppliers. There are several ways of examining a company which a stakeholder may utilise, including:

- a review and an assessment of the company's annual reports and financial accounts and statements
- auditor's endorsement of the annul report and accounts or lack of report
- credit rating agency's assessment
- ratio analysis of the financial accounts against benchmark peers
- market analyst's report on the company in its competition and market context
- a technical analysis of the company's performance over time.

A study over several years is, perhaps, more useful than the last reported period. Such a study will seek trends and reasons for changes, eg it was reported that the credit rating of some universities might have been affected by the performance of their private finance initiative construction schemes which had been delayed.

Accounting ratios and other ratios

Accounting ratios may be used as a form of benchmarking. They enable comparisons as follows:

- to compare a company's progress from accounting period to accounting period, eg every year
- to compare overall performance of two or more companies in the same field over a series of time period
- where detailed enough, highlight instances of exceptional management operational performance or vice versa.

The data used may be from published financial statements or from internal financial accounting and management accounting documentation. Internal material may be used to analyse a division or a subsidiary: external material is more limited.

A very large number of ratios are used for a full analysis of a company's performance by external study, ie from both the profit and loss account and from the balance sheet (as published as interim or annual reports). Of course, the type of business that the company conducts will determine the appropriateness of the ratios. The ratios given in the Box 23.2 are a very limited sample of those which may suit someone who is looking at a property company.

Box 23.2 A selection of typical ratios from the profit and loss account and from the balance sheet

	Profit and account ratios	
Price/earnings ratio	$\dfrac{\text{price of the share}}{\text{earnings per share}}$	A comparator for companies in the same sector
Turnover/assets	$\dfrac{\text{turnover or sales}}{\text{net assets}}$	Shows how quickly the company is turning its assets
	Balance sheet ratios	
Net asset value/share	$\dfrac{\text{assets less debt}}{\text{total number of shares}}$	Indicates the worth of the shares in assets — compare it with the share price
Return on capital	$\dfrac{\text{net income}}{\text{net assets}}$	Compare the result with the cost of capital
Gearing	$\dfrac{\text{debt}}{\text{equity}}$	Shows the proportion of debt to equity In a growing company, the return on equity grows more quickly
Liquid ratio	$\dfrac{\text{liquid assets}}{\text{current liabilities}}$	Shows how easily the company could pay immediate debts

---

It should be borne in mind that the accounts of local authorities and other public differ from those of private sector entities and that every sector or sub-sector has its own peculiarities.

The Centre for Interfirm Comparison offers a number of services based on benchmarking and the use of ratios. It does studies based on particular industries, professions and so on. It has covered a very large number of groups of organisations, including:

- solicitors
- chartered surveyors
- residential property investors
- fund raising by charities.

Developer performance

The developer will want to be assured of performance on two levels, namely:

- the procurement of the development should meet the specifications and performance indicators for value, time, cost and quality (physical), eg were snagging incidents low
- the objectives of ownership should be met, eg for the investor developer is the rental level expected achieved.

Development ratios are used to assess a project, including:

- development profit ratios — profit:development costs; profit:gross development value
- building cost ratios — cost:floor area
- investment ratios — internal rate of return:cost of finance.

In practice, the client should seek an explanation of the approach to such analysis used in a particular project. For instance in a one-off development, where the funds used are entirely a single source of debt the internal rate of return could be compared to the cost of borrowing. In another instance, a company finance officer may require that the "weighted cost of funding the company" be used. On the other hand, when considering two prospective alternative projects their estimated internal rate of returns may be compared.

Investor performance

An investor in property will want an annual net rental return and hope to gain an increased price when the property is realised at some unknown time in the future. There are a number of approaches to measuring the performance of the investment, including:

- the return on the capital that is invested
- the total return on the investment.

However, the investor will probably want to benchmark the investment in some way. This may be done by considering one or more of the following:

- the return that could be achieved by investing in government bonds
- the cost of raising funds to buy the investment
- the return on alternative property (and other) investments, bearing in mind the relative risks of each alternative proposal.

Dealer performance

For each property project, the dealer will aim to make a profit. The profit will be expected to cover the following:

- interest on the capital employed in the business — the capital will be property held, plant and machinery, materials, work-in-progress and working capital (cash)
- risk faced on the project
- remuneration for the time spent working on the project
- taxation on income from the project, ie on the profit.

Project performance

There are many stakeholders for every development project. No doubt, some will be winners, others losers and yet others will feel that they are evens. There is no one overall measure for a development. For instance, management's performance on a development project is, in essence, four dimensional. Does the project meet the developer's expectations on parameters concerned with quality, time, value and cost? Other stakeholders will have other measures or performance indicators. For this reason Box 23.2 briefly sets out the various stakeholders and the quantitive or qualatitive perspectives which they might use.

Box 23.2 Stakeholders' perspectives of a development

Client	• value, cost, time and quality • fit to objectives of ownership
Professionals	• job well done • client satisfaction
Contractor	• ease of working relationships • absence of, or ease of any resolving, disputes
Occupier	• suitability for use, operating costs, etc
Neighbouring • owners • occupiers	• impact on value of one's own property • nuisance caused by physical factors — noise fumes, smoke, deposits of soot etc, artificial light, vibration, etc • increase or loss of amenity, eg increased traffic congestion
Local authority	• compliance with standards for sustainability and acceptability (see Chapters 13 and 14) • impact on employment, rates and voters' expectations etc
Funder	• amount of owner's equity • on default, ease of liquidity and recovery of what is due

---

Taxation and performance

Taxation is an important variable in development situations which the developer and other parties must have regard to in deciding between options for a scheme. The client will usually wish to be advised of the best net of tax outcome of any range of options. It is unlikely that taxation will be the main determinant for the client but it may be important in reaching a decision.

Gross yield

A straight forward approach to measuring the performance of different investments or property proposals is to consider their gross yields, ie before income tax (or other tax) is deducted. Gross yield comparisons have the merit that common yields are quoted in this way, eg building society interest is quoted gross.

Net net basis or true net basis

Where an appraisal is being undertaken for a particular person a gross yield approach offers insights. However, a more subjective or client-oriented approach would be to consider each investment or option the person could make on the basis of true net outcomes, ie after the burden (or otherwise) of taxation has been considered and allowed for in the calculations.

Performance in the public sector

Many of the points looked at above will relate to the public and voluntary sectors. Nevertheless, there are differences in perspective. So, the appraisals used above have their place in the public sector but the government has drawn up the best value approach to the assessment and review of the way in which local government and other public bodies deliver their services to the community.

Comprehensive performance assessment

The Local Government Act 1999 provides for best value performance measurement.

The Audit Commission conducts comprehensive performance assessments of local councils and other public bodies, with emphasis on community matters, eg housing performance. It classifies the performance under quality of service, eg "good", and likelihood of improvement, eg "no", (see Box 23.3).

Box 23.3 Comprehensive performance review — performance classification

Quality of service		Likelihood of improvement
Excellent	(3 stars)	Yes
Good	(2 stars)	Likely
Fair	(1 star)	Unlikely
Poor	(0 star)	No

The Audit Commission's approach is to select one function, eg legal services, housing management or building control. It has indicated a seven step procedure for comprehensive performance assessment. However, Box 23.13 shows schematically an approach to a CPA which extends the detail to 10 steps.

Box 23.4 Schematic comprehensive performance assessment of a service of, say, a local authority

Step	Local authority	Other	Inspectors
1	• receives notification of forthcoming CPA • prepares self assessment		
2		• peer challenge by consultancy group with independent members	
3	• prepares document inspection pack • sends it to the inspection team		
4			• receives documentation pack • team familiarises itself with the material
5a	• greets team • accommodates them with office and telephones etc		
5b			• visits council • undertakes inspection • does reality checks • prepares draft interim report
6a		• team reviews draft with other Audit Commission inspectors	
6b			• finalises the interim report • sends copy to LA preview team for interim challenge
7	• local authority pre-view officers comment back on interim challenge		• prepares final report
8	• receives the presentation		• presents the final report to the council and senior staff
9	• inform staff of the report • reviews the findings • develops initiatives to address weaknesses • seeks to improve performance overall		
10	• implements and monitors initiative		

Valuations and Appraisals

Aim

To explain how property is valued and developments (and other situations) are appraised

Objectives

- **to describe the role and activities of the valuer or appraiser**
- **to briefly explain the use of different methods of valuation and appraisal**

Introduction

A valuation or appraisal is at the heart of many of the property decisions that owners and occupiers make when they buy, sell, improve or develop property. Valuations are usually made to open market value at a particular date in the sense they are objective and not related to the personal circumstances of the owner-seller or the prospective buyer. However, it is important for the user of the various techniques to appreciate a number of features, namely:

- before commencing, to consider what a technique is being used
- the limitations of each application of a particular technique
- what the output(s) of each application represents
- in some instances, the use of sensitivity to explore different situations (by changing the assumptions about a variable or a combination of variables).

Where a property is to be purchased by raising money, the buyer may have several possible sources of loan funds, each of which has different terms and conditions. Furthermore the choice of finance may not be restricted to a loan but raising equity capital may be possible by issuing shares. An appraisal may be needed to decide which method of financing should be used, with any taxation implications built into the various analyses. In these circumstances, the prospective buyer may require:

- a valuation as above, ie objective assessment to determine the price which may have to be paid

- one or more appraisals with a subjective bias to find out whether it is possible to pay and, if so, which fund raising approach should be used.

The methods of valuation are explained in this chapter (with some reference to Chapter 30). In addition, some illustrative appraisals are included to give a flavour of the approaches.

Value or price?

The value of a property is an estimate of price while a price is the reality of a transaction, ie the actual consideration that was paid for a property by the purchaser. A value is, therefore, a person's opinion of the price at a particular date reached by the application of methodology by a knowledgeable and experienced individual.

Determination of price

A property's price is determined or fixed by one of a number of methods. Box 24.1 sets out the approaches. These are, of course, an element of the methods of transferring property explored in Chapter 12.

Professional approach

In their valuation and appraisal work, valuers tend to use four principal traditional methods of valuation and several discounted cash flow approaches.

The four principal methods of valuation or a combination of methods may be used to find the open market value of property, either to capital value or to rental value. Generally, there is no prescribed single approach to each method, although the prudent professional valuer will normally have regard to relevant items from the following:

- any statutory provisions governing the approach, eg definitions and assumptions
- any mandatory standard set or recognised by a professional body
- any discretionary guidance given or recognised by a professional body, eg the RICS Code of Measuring Practice
- any Lands Tribunal or other court's determination on the particular topic being considered.

A member of a professional body is obliged to follow mandatory standards and would be well advised to follow guidance; any departure from the guidance needs to be justified, if necessary, in court or at any hearing conducted by the professional body.

Appraisal and valuation standards

For certain types of valuation, valuers work to standards laid down by relevant international or national bodies. In the UK the RICS Appraisal and Valuation Standards of May 2003 are required for the valuation of company assets for the financial accounts of a company or for a listing on a stock exchange.

Box 24.1 Approaches to determination of prices in property transactions

Private treaty negotiation	• agreeing the methodology, assumptions, measurements and other factors to be used in valuation • seeking comparables and other value data • valuing the properties independently • discussing their results and reaching agreement • lack of agreement means there is no price and no transaction
Auction	• particulars contain a draft contract for the sale • competitive bidding • in an open or closed forum • successful bid creates a contract • bidder must be prepared to follow with deposit, signed contract and, in due course, the balance of the funds
Arbitration	• the parties put their cases to the arbitrator • the arbitrator determines the price
Independent expert	• may hear evidence and use own experience etc to make decision
Negotiation under a code	• as for the private treaty negotiation, but the parties follow a pre-set code to value, eg the rules and assumptions of compulsory purchase
Management decision or formula	• management determine an approach, eg for internal appropriation, use a prescribed discount rate • formula, eg cost plus allowance for risk and remuneration
Scale	• price scale for wayleaves and easements • price for service is fixed on pre-set formula, eg the Ryde Scale for professional fees
Surrogate price	• price for new type of property is proportion of price of another type of property • at least until a market develops for the property

Of course, if a client company's professional financial representatives require the valuations to comply with other standards, eg the European Valuation Standards (EVA) 2003, the valuer would accept the instructions. In fact the valuer would be obliged to comply with any relevant mandatory standards imposed by the London Stock Exchange or other regulatory bodies.

There has been a gradual developing consensus on valuation standards at an international level so that valuation work is becoming somewhat similar in many countries.

Standards and measurement

The fifth edition of the RICS Code of Measuring Practice provides the professional valuer with an illustrated guide of definitions and issues which must be addressed when particular properties are to be measured. It gives a number of questions which the prudent valuer will think though when planning a measurement job. The code also refers to the need for the valuer to consider the impact of any statutory obligations, eg the Property Misdescriptions Act 1991, inherent in the work situation.

Building works are measured according to the Standard Method of Measurement. Generally, unless the valuer has a good estimate of specific construction works, some estimate will be made from experience of like buildings, otherwise sources, such as the RICS Building Cost Information Service will be used.

Information

A busy valuation practice is monitoring the catchment area for information which can be analysed to provide a justification or confirmation (or otherwise) of a valuer's views on capital or rental trends or a recognition of sharp changes in, or a flatness of the market. First hand transactions, particularly with negotiation, attendance at auctions and a reading of material with up to date hard evidence of prices and yields will be utilised.

Traditional methods of valuation

The four basic methods of valuation are:

- comparative
- investment
- residual
- contractor's basis.

A fifth method, used in rating, is known as the formula method but is becoming less used than hitherto.

Comparative method

The approach to the comparative method of valuation is essentially an analysis of transactions in similar properties with similar characteristics, eg freeholds with one vacant or let at full rents, so that the analysis results in capital or rental values of land or of buildings being expressed in units of price to area. The results are then applied to the areas of the subject property. Generally, both rental values and capital values may be obtained by using the method

The types of property which may be valued by the comparative method include:

- agricultural land, sports fields and other "space" property
- shops, offices and industrial floor space
- houses and flats.

Care needs to be taken that the appropriate method of measurement is used, ie observance of the RICS Code of Measuring Practice should ensure that when negotiations commence both parties are speaking the same language.

Investment method

Investment properties are bought for the yield or return which they give. Yield is usually expressed as a percentage of the capital value. Briefly, a vacant property may be valued by converting the estimated net rental open market value into an equivalent capital sum. The conversion is done by multiplying it by a number (known as the year's purchase — the reciprocal of the yield). The figures are obtained by using the comparison method of analysing the rents or prices of recent transactions where properties have been let or sold. The rental value and yield are thus obtained. Even an owner occupied property (a shop, say) has a *notional yield* — the owner could vacate and let at the market rent.

The basic equations for freehold valuations are:

$$\frac{\text{Net rental value}}{\text{capital value}} \times 100 = \text{yield \%}$$

$$\text{Net rental value} \times \frac{100}{\text{yield}} = \text{capital value}$$

$$\frac{100}{\text{yield}} = \text{years' purchase (YP)}$$

The pattern of yields in the market is always changing but an indication is given in Box 24.2.

Box 24.2 Indicative pattern of yields for different types of freehold property

Retail property	4.5 to 7.5%
Offices	5.0 to 8.0%
Industrial	8.0 to 10%
Warehousing	8.5 to 11%
Leisure	7.0 to 10%

Sales of recently let freehold property are analysed to find the yield on the investment. In the example below, the yield obtained is 5% and the years' purchase is 20 YP. With experience over a period, an availability of transactions and knowledge of the local economy, etc, a valuer is able to read yields in the property market. At any one time there is a pattern of interrelated yields in the local market. Thus a single change in market conditions will, say, affect the pattern of yields. Hopefully, a local valuer will know of this and endeavour to appreciate the underlying pressures. For example, the recent increases in the rate of interest by the Bank of England's Monetary Policy Committee have pushed up building society lending rates. The consequences are that;

- house buyers are holding off the market or reducing their offers
- existing owner borrowers are receiving higher bills for interest on their variable rate mortgages
- some owner borrowers, particularly those at the margin of being able to cope, are trying to down-market by selling their homes
- others at the margin who cannot find a buyer or will not sell because they reckon they will manage, are borrowing from sub-prime lenders (at higher rates of interest)

- some others, who cannot borrow from anyone, cannot pay the arrears of interest and principal — so the building societies or other lenders seek repossessions.

As a result prices continue to drop until the whole economy is where the monetary committee begin to reduce the rate. In effect, the *notional yield* implicit in an owner occupied house purchase (for a capital sum) has increased as prices fall.

Residual approach

The residual approach is used to value building land, ie land with planning permission or the prospect of the developer obtaining planning permission. The basic formula is takes the estimated market value of the completed property — the gross development value and take from it all the development costs which it is estimated will be incurred on creating the building together with an allowance for risk and profit:

Box 24.3 Residual valuation in outline

		£
	Gross development value	
Less	Development costs	
	Risk and profit	_____
	Equals: Land value =	_____

The following properties may be valued using the residual method:

- existing property capable of conversion, with an appropriate contingency allowance for the unexpected
- greenfield land with planning permission
- brownfield land with planning permission, with an allowance for works
- contaminated land, with allowances for remedial works and for stigma
- land with hope value.

Uncertainties in the last three valuations create problems for the valuer. In a number of cases the Lands Tribunal has criticised the use of the residual method however it is often used as a supporting approach to the comparison method.

Accounts or profits method

The profits method is used for hotels, restaurants, leisure facilities and other unusual properties where there is less market information available to use the other approaches.

Box 24.4 Profits or accounts method of valuation to find rent

		£
	Turnover	
Less	Stock sold or consumed	
Adjust	+ or – Stock change	_____
	Gross trading profit =	
Less	Working expenses (excluding rent paid)	_____
	Net trading profit =	
Less	Interest on capital Risk and remuneration	_____
	Rental value =	_____

Notes:
The adjustment for stock allows for any increase or decrease in stock in the stockroom.
The rent paid, if any, is excluded because the valuer is valuing for rent.
(A similar calculation is used to value goodwill — see Chapter 23.)

Contractor's basis

The cost basis or contractor's basis uses the cost of the land and the construction costs to arrive at a capital value. The method follows the format given here:

Box 24.5 Contractor's basis of valuation in outline

		£
	Cost of land	
Plus:	Cost of building	
Less:	Allowance	_____
Equals:	Value of Property	_____

The method is commonly used where there is no market for the property, eg church buildings, or for appropriations within or between public bodies, eg schools, hospitals. It is commonly used for these properties to find rental value for non-domestic rating.

Valuation of difficult property

Valuations involving certain kinds of property are likely to present the valuer with particular problems. Such property includes:

- brown field sites
- land which is known or thought to be contaminated land
- land with hope value
- landlocked sites
- listed buildings
- other old property in general
- property where there is the prospect of marriage value.

Brown field sites

A brown field site is one which has been in developed in the past and is now available for redevelopment. They tend to have foreseen and, possibly, unforeseen problems which the valuer must allow for in any appraisal.

Land with past contaminative uses

Land for development which is contaminated or is likely to be contaminated raises a number of issues for the valuer, such as:

- the investigation and remediation will need to be paid for
- the polluter pays principle operates but only where the polluter is known and is able to pay
- the kind of development which would be permitted and the cost of remediation which would allow the said development
- the effectiveness of any remediation may be uncertain
- any existing insurances, guarantees or warrantees may be or may become invalid
- after remediation it may still be difficult to obtain finance for the development
- the insurer may impose terms and conditions which are, as yet, unknown.

Land with hope value

A value or price for land with an element of hope value is one which reflects the market's expectation that the land will enjoy a future planning permission for development.

Landlocked sites

Under compulsory purchase a number of cases have involved landlocked sites, ie land without adequate access to the highway for the kind of development envisaged. Various approaches to the valuations were used and almost all involved the assumed acquisition of access land. In valuing the back land, the owner of the access land may have argued for and received a proportion of the development value. Of course, if there are several prospective access routes, it may be possible to secure access more cheaply than would be possible where there is only one.

Box 24.6 Valuation of landlocked development site — marriage value outlined

		£
	Gross development value (of combined site)	G
Less	Costs of development	(C)
	Risk and profit	(P)
	Land value (with access) =	G – (C+P) = X
Less	Value of site (without access)	(E)
	Development value =	X – E
Less	Value of the site for access	(A)
Hence	Share for access land, say @ 30 % =	0.3 (X – (E+A)) = Y
Open market value of access land =		Y + A

Notes
The share will depend on the availability of other possible access routes.
The owner of access land will want its existing value plus a share of the development value.

Listed buildings

There is a presumption against the demolition of a listed building. Most development works including demolition require listed building consent as well as planning permission, if necessary. Such consents tend to take time and where works are permitted, the cost of the works tends to be substantially higher than those in an unlisted building.

Old property in general

In general, older property tends to have problems (Box 24.7).

The valuer may be required to allow for matters like these in the valuation. Of course, the nature of the valuation may enable them to be ignored completely. Thus, a different approach where

- a residual valuation assuming there is redevelopment after demolition and clearance of the site, or
- an insurance valuation on the assumption that the building would be replaced with a modern substitute building different approach.

Box 24.7 Typical characteristics to be found in old buildings

Lack of modern facilities	• installation and adaptation may be costly
Inadequate space	• may not provide for sufficient parking and other planning standards
Old structure and fabric	• may not allow easy modernisation, eg for noise insulation
Out-of-date or old worn services and plant and machinery	• removal and renewal may be costly • obtaining parts may be costly or impossible • complete renewal may be required, eg electrical system
Asbestos	• requiring special treatment in removing it and monitoring
Inadequate access and other facilities	• for persons with disabilities • works in accord with the Disability Discrimination Act 1995 required
Structural weaknesses	• works required to strengthen the building
Vermin infestation	• removal may require costly specialist treatment
Listing	• listed building consent required for any demolition or works • labour tends to be in trades which are costly • materials and components tend to be costly

---

Marriage value

Marriage value for a landlocked property has been considered above. However, it may arise where any buyer who owns an interest in a property, eg a leaseholder, wishes to buy the interest of another, eg the freeholder's reversionary interest. Normally, marriage value may be negotiated in the price. Although, where the purchase arises under statute marriage value may be excluded under the Act.

Appraisal techniques

In Chapter 23, financial and valuation appraisals are explained in the context of the measurement of performance; the topics dealt with in this section puts the appraisals in an operational context.

Development budgets

A development budget will be used by a developer to ascertain whether, for example, a dealing operation will be profitable or not. The development budget approach will show:

- the accumulation of expenditure (and interest) until receipts catch-up
- whether the break-even point of the project is going to be achieved
- the eventual profit on the project (or loss).

Essentially, the expenditures and receipts are portrayed in a grid display depicting, say, monthly or quarterly periods. In the early months the net amount is likely to be negative; as sales are achieved the successful project will move into profit, ie all accumulated expenditure, including interest, has been covered by accumulated turnover.

Budgetary control

The development budget has just been described in the context of an appraisal. As the project progresses it is likely that some form of the approach will be used repeatedly as a management technique to monitor progress and to alter operational plans for construction, promotion of the project's units and sales.

Development appraisals

A residual valuation may be used to appraise a proposed development or a development which has commenced. The three basic equations for this are shown in Box 24.8.

Box 24.8 Development appraisals

£

	Gross development value	
Less	Development costs	
	Estimated risk and profit	————
	Land value =	————

£

	Gross development value	
Less	Development costs	
	Actual land cost	————
	Profit =	————

£

	Gross development value	
Less	Actual land cost	
	Building costs	————
	Estimated profit =	————

The first equation shows a valuer's traditional residual valuation. Estimates of the factors on the right hand side are used to value a site with planning permission.

The other two equations show outcomes where the appraisals are done:

- after the land has been bought
- the GDV may be estimated or be the sale price of the building on a forward sale
- the development costs may be a mix of estimates or fixed prices, eg a fixed price contract for the building contract.

Discounted cash flow techniques

Relatively complex appraisals may be undertaken using one or more of several discounted cash flow techniques, such as:

- net present value
- internal rate of return.

Discounted cash flow appraisals are used with a more specific regard to timing or programming of receipts and expenditures, ie it may portray monthly, quarterly or yearly periods (whereas the residual valuation is generally much less precise). They have been used to show many aspects of development, construction and investment, including:

- the effects of taxation
- the impact of differing wage rates
- effects of differing rates of inflation
- the way in which different components may affect the long term life-cycle costs.

The approach is similar to that for the development budget except that interest is not included in the grid as a periodic expenditure. Instead a discount rate is applied to each year's net cash flow to calculate the net present value of the project.

The output of such a calculation depends on what the appraiser wants to find out. Thus, the net present value will be:

- land value — where the land element of the costs are not included (but an allowance for risk and profit is included) or
- profit — where the land cost is included but there is no allowance for risk and profit.

Funding Property Transactions and Works

25

Aim

To explain how the buying of property and developing land, etc are funded

Objectives

- **to describe the different needs in funding property**
- **to classify expenditure and indicate how it is raised**
- **to describe corporate funding and project funding**
- **to show how infrastructure projects are funded**
- **to explain how regeneration projects are funded operate**

Introduction

This chapter looks at finance for buyers, developers and others in the private sector: Chapter 33 is concerned with those who need funds for what are essentially public sector and voluntary sector projects.

A look at the classification of expenditure and the need for funds are the first two main topics of this chapter. These are followed by an account of the sources of money and the various ways in which it is supplied to owners, developers and others. This provides a brief review of those involved in the money market for property, including an explanation of the illustrative budgets which may be used to support an application for funds.

Ways of raising funds is examined from the perspectives of the organisation and the project; essentially corporate funding and project funding respectively. The last two sections cover aspects of taxation and insurances of those involved in funding property proposals.

Classification of expenditure

Without funds individuals and organisations cannot run or operate a property and they cannot proceed with purchases of land and buildings or with the development of projects. Thus, money is needed by the owner both to buy or create a property and to run or operate it.

For owner occupiers and investors, buying or creating a property involves an outlay of money which is regarded as "capital" (the property is a capital asset). The running or operating expenses are "revenue" items. As far as the property dealer is concerned, the technical perspective is that the property is stock-in-trade and any expenses are of a revenue nature when calculating profit. The distinction is particularly important for taxation.

The need for funds

All owners will require capital funds to buy or develop property but, apart from dealers, property owners need a fund to run or operate a building and any grounds held with it. Since dealers are selling on or developing and then selling, they are only likely to need funds to pay for the security, insurance and relatively few items involved in running a shut down or closed property awaiting development or sale.

Annual outlays

Maintaining and operating a property or estate may require the owner or occupier to spend money on such items as:

- repairs and maintenance
- insurances
- services, including gardening, heating, lifts, lighting and porterage
- rates and taxes
- management and supervision
- rent (where the owner or occupier is a tenant)
- interest on borrowed money.

Where a property is let, the precise wording of each lease will determine the division of outlays between the landlord and the one or more tenants (see Chapter 3).

Payments out of income

Resident owners and business owners aim to pay for these outlays from income, ie salaries, business profits, rents or sources of income. Of course, where any tenants pay a service charge the owners will pay from it. Dealers may do the same, but only dealers might otherwise aim to pay any such annual outlays from borrowed funds (or other capital sources).

Deficit financing and bankruptcy

Property owners will aim to at least break-even but where there is no income from the property or the outlays are greater than income, the owner must make up the difference by deficit financing, ie financing from other sources. Any inability to do that is likely to result in the sale of the property to pay off creditors, including any mortgagees. If the proceeds of sale do not meet the debts, bearing in mind the hierarchy of creditors, the owner faces bankruptcy.

Other sources of money

As stressed, the need for finance is imperative. It follows that other sources may need to be sought, including:

- the realisation of other property by sale or part exchange — real property, shares and so on
- borrowing short, medium or long term debt
- a flotation or the issuing of shares in an existing holding company
- the renegotiating or rescheduling of existing debt to reduce annual payments.

Financial control

It follows that the start of any prospective purchase of a property or any development proposal must follow a careful risk appraisal, financial appraisal and the preparation of a budget! During the development of a project, the owner will need to ensure careful budgetary control.

Funding the organisation

This section examines how funds may be raised to create organisations (rather than a project). The funding of government sponsored public bodies is dealt with in Chapter 33. However, a project under the private finance initiative might be regarded as a hybrid body. Although it is covered in Chapter 33 the devices commonly used to raise funds from the private sector are covered in this chapter.

Finance for a new business

An individual or small group may found a business with their own capital and, if necessary, with capital borrowed from a bank, finance house, relatives or friends. The borrowed funds are likely to be secured on the personal assets of the founders of the business. Typically, an individual may begin the business in their own name or as a sole trader with a business name, generally with unlimited liability, ie to the extent that it, the liability, cannot be transferred by, say, insurance. An unincorporated group will trade as a partnership with joint and several liabilities.

Thus, the individual or the group may create a private limited company. The step to incorporation is taken for issues that the founder of a business needs to really address from the start, such as:

- possible access to outside funds for the future development of the business, eg by the issue of shares for cash
- attaining limited liability for the directors, so protecting personal wealth and the security of their families
- the mixes of future capital taxation and income taxation to be faced by the business organisation and the individual founders
- succession to the leadership role in the business.

As a sole trader or as a private limited company it may be difficult to repeatedly raise finance from the original sources or local sources. Indeed, some of lenders or shareholders may want to be repaid their capital after a relatively short period.

Finance for an established company

For the private company, several courses are open. Not all will necessarily involve equity funds. The opportunities include:

- a trade sale, but this may result in the owners staying in business but losing control or totally withdrawing from the business
- an equity capital injection by a venture capital company with the view to a future flotation — the board may be strengthened by representation from the venture capital company
- a flotation on a stock exchange — often the next step for a company which cannot fund itself with organic cash-generative trading.

The AIM or the Ofex exchanges are likely exchanges for a relatively small company or one with a trading history which does not fit the requirements of the London Stock Exchange. Both markets are somewhat less costly than the main market. At this stage the company has become a public limited company and the scope for raising funds form the investment market is, in general, available.

Investment devices for raising funds

A number of devices for raising funds are available to the board of a company, including;

- further issues of ordinary shares
- rights issue of ordinary shares
- dividend re-investment plan (DRIP — at least DRIP is not paying out funds)
- debenture stock
- preference shares.

Shares may be issued for cash by placing them with institutions or by offering them to existing and new shareholders. A rights issue is a similar exercise but must be offered to the existing shareholders, invariably at a discount to the share's market price. A company wishing to acquire another company or, say, an estate (or a single property) may be able to use ordinary shares as consideration.

DRIP is an issue of shares to the existing shareholders in lieu of their interim or final dividend — the shareholders sign up, and sign off when they wish to do so. (For the shareholders this follows the accumulation principle of reinvesting the return. For the company it reduces the outflow of cash.)

The other devices are in effect loans to the company in return for interest. Sometimes a debenture or preference shares may be converted into ordinary shares within a prescribed period.

As far as the board and shareholders are concerned the following should be considered:

- the impact on the share price which any device may cause
- the relative cost to the company of each of the alternative ways of raising funds
- the differences, if any, to the company and other stakeholders in the taxation treatment of the different devices.

Sources of funds

Flotation on the Stock Exchange

In recent years the open share flotation, ie to the general investing public, as almost dried up. Financial advisers to companies find, in general, that it is better to place the shares with financial institutions and other intermediaries. The reasons are:

- the shareholdings are more substantial and less costly to service thereafter
- the time taken for an initial public offer by placing is shorter than an open offer, ie the placing is shorter
- the prospectus is less voluminous (because the information is going to expert investors) and less in number need be produced
- the cost of the IPO is cheaper.

Bonds and notes

Money put into bonds and notes by investors is raised by companies and others. The bonds and notes on the financial market are bought for an interest rate return, the coupon, for periods of more than one year and, possibly, as much as, say, 30 years, or even for an indefinite period. The principal bonds are corporate bonds and Eurobonds. However, developers may seek to issue a deep discount bond where the coupon is substantially less (even zero) than the norm coupon for the term envisaged.

The purpose of this bond is to reduce interest outflow when a development project is not producing a rental flow or a rental flow which is much less than on completion of the letting programme. Sometimes the bond's coupon is stepped to a higher level, say +1% every three years, from the rate specified for the first three years. Of course, the quid pro quo is that the face value of the bond (the amount to be paid back) is greater than the sum that the developer receives when the bond is placed. In some ways the deep discount bond has the flavour of project funding rather than corporate finance.

Funding the project

Raising funds, other than by corporate funding devices, is sometimes termed project funding. It includes:

- mortgages — loans secured on existing property or property to be acquired
- sale and leaseback — the owner sells an asset or an estate of properties and leases them back form the new owner
- reversion equity release (sale) — the home owner sell the dwelling for a regular income for life, retaining the right to remain in the property
- equity release loan (loan) — the home owner borrows a proportion of the open market value of the dwelling, retains possession and pays interest on a rolled-up basis
- franchise — the owner of a business creates a model of the business which enables it to be replicated and operated by independent but tied entrepreneurs
- securitisation — the owner of a property or a portfolio of loans bundles them and create a security which others buy for the return (backed by the rents or the interest).

Generally, for the borrower loan stock bearing interest is cheaper way of raising money than securitising assets where the income from the assets back or secure the units the lenders receive. It seems that Rail Network is raising £30bn with loan stock rather than securitising its revenue streams from the train operating companies.

Mortgages

A mortgage is a loan where property is offered as security for it by the property owner. It is the common way of financing home ownership or the ownership of business or investment property. The lender's approach is minimise risk. A buyer of a house will find that the availability of a loan is based upon such principles of lending as:

- a norm for the loan to value ratio of less than, say 70% but higher percentages are not uncommon
- a maximum loan of, say, up to five times net income but for couples joint incomes may be considered
- interest to net income ratio is less than one, ie for let property
- where the property is leasehold, the duration of the loan is shorter than the lease by, say 30 years
- the buyer carries out works to the property, ie to improve it (as necessary and put into a good condition).

If the borrower defaults the lender has a number of remedies, including repossession and sale. Thus, as the Monetary Policy Committee of the Bank of England increases the interest rate, the number of repossessions tends to increase. The loan to value ratio should, however, ensure that the lender covers the outstanding principal, the arrears of interest and the costs of the sale.

Banks and building societies

The banks and building societies offer mortgages and other financial products to house buyers and businesses. Generally, they operate to codes, the Banking Code, the Business Banking Code and the Mortgage Code respectively. The codes set out the relationships between the parties in terms of fairness and best commercial practice. The Banking Code Standards Board oversees compliance with the two banking codes and is empowered to enforce them and take action against an errant bank.

Ombudsman for financial services covers the field but it is only recently that mortgages have come within the purview of the Financial Services Authority.

Equity release

With the rise of house prices in recent years, many older home owners have been able to release some of the value by one of the two basic forms of equity release. The prudent owner will carefully weigh up the alternative approaches and, having made a choice, carefully considers which company offers the most suitable scheme for the approach chosen. Box 25.1 compares the loan approach with the sale approach.

The owner's main concern is likely to be whether negative equity will arise. Many schemes provide that it will not do so.

Box 25.1 The two main approaches to equity release

Feature	Mortgage for life (loan) approach	Reversion (sale) approach
Ownership	• retained until sale during life, eg going to a care home or on death (by the administrator • may cover two lives (spouses)	• passes to the company or shared
Possession	• retained by the owner	• until death
Proportion of value	• a percentage which varies with age	• all or a proportion • at a discounted price
Further funds	• yes, in small percentages	• not if all the equity was sold
Repayment of the amount raised	• yes, property sold on death • possibly, if a life time sale*	• no (outright sale originally)
Shifting funds to another property	• company may permit as a policy • if so, second property must be acceptable security	• owner may use original money
Negative equity (house prices drop)	• if principal and interest exceeds value on death, agreement will specify limit, if any • important issue to be checked before contracting	• no (outright sale originally)
Heirs' inheritance	• balance of sale price on death less what is owed (plus balance of loan in estate) • zero could be lower limit • negative equity possible but owner could avoid possibility	• none (house sold) • balance of sale price in the estate
Costs	• survey fee payable outright • costs and solicitor's fees in the loan amount	• payable by the owner up front

Protection

From 1 November 2004 mortgages and, in this context lifetime mortgages (unlike reversion plans), will come under the purview of the Financial Services Authority.

Also, Safe Home Income Plans (SHIP) is an association of companies offering this kind business; its members subscribe to a code of customer protection.

Franchise

For the lessors (or sellers) of retail property the franchise market is a niche where the quality of the prospective tenant's franchise business and the use and operations thereof are normally transparent

— the business-to-be is replicated or "cloned" elsewhere by other franchisees. Furthermore, the business is backed and supported by the experienced franchisor and, in most cases, a team of business advisors, sources of finance, ie the franchisor has a strong marketing mix. Box 25.2 shows the marketing mix of a typical franchised roadside eatery and lodgings.

Box 25.2 Marketing mix for a typical franchised roadside eatery and lodging (for franchisee)

Location	• sometimes off a motorway • often on a trunk road • visual or signage
Image	• recognisable franchise image, colour etc. • logo visible
Appearance	• building style and décor may be common to franchise
Car parking, petrol etc	• car parking is provided — owner can usually see car • petrol filling services sometimes on the same site
Accommodation for dining	• interior design and layout recognisable for the franchise • common style of furniture, menus and staff dress code etc • payment systems common, eg credit card
Service	• menus, waiting style etc. is common to franchise • self service or staff is common • menus are common and prepared in the same way
Retail services	• usually simple travellers' requirements, eg sweets, cigarettes, etc
Rest rooms	• franchisor provides timetable for cleaning and standards
Lodging accommodation	• if any, basic and comfortable • family room charge is common • meals may be in another building • en suite bathroom
Retail services	• in reception, basic overnight and travellers' fare • water and soft drinks — vending machine

---

Although franchise is not in itself a source of finance, the opportunity-cost for the franchisor is to expand the business by an owned chain of retail businesses. By having franchisees own and run the unit-businesses the franchisor is, in effect financing the business expansion on the outside.

Funding private finance initiative (PFI) projects

In the public sector numerous private finance initiative property development projects, eg schools and university student accommodation, as well as projects for complete services, eg information technology, have been undertaken. Many were funded from service sources or by contractors' funds (where they made an equity investment), but financial institutions, eg banks, have provided the finance, often by syndicated loans. There is now a secondary market in private finance initiative interests.

Islamic finance

Sharia

Financial transactions which comply with Islamic law, Sharia, have become more common in recent years. The essence of a compliant transaction is that it reflects the following:

- placing funds in an account does not involve the receipt of riba or interest
- receiving funds does not require the payment of interest
- the benefit or cost in such transactions is profit (not riba) to the payer or receiver of funds respectively — there is a sharing of the risk
- the funds placed with the recipient bank or financial institution must not be used for non-compliant purposes, ie placed with businesses or organisations involved in alcohol, gambling, prostitution or other proscribed activities.

Institutions offering Islamic instruments

There are now a limited number of UK banks which offer customers ways of placing and raising funds which do not involve the receipt or payment respectively of interest or riba. For instance, since 1998 Barclays Bank plc has offered an Islamic instrument for house purchase. In October 2004 there was the flotation of a company on the AIM with the purpose of offering Sharia compliant financial business to its customers with a variety of financial instruments.

Compliance

Any financial institution which offers Islamic instruments to customers must demonstrate compliance with Islamic law by opening its financial instruments to a recognised authority on Islamic financial law for scrutiny and approval. The compliance authorisation will normally be internal as a first step and then external.

Islamic mortgages

An Islamic mortgage is a transaction where those wanting to have, say, a home, are leaseholders until it has been paid for by rent. The mortgage is arranged as follows:

- the financial institution buys the property on behalf of the home-seeker
- the home-seeker moves in and pays rent for a pre-determined period
- when the last payment of rent is made, the house is transferred to the "tenant" who becomes the owner.

There was some concern that such financial instruments would not be attractive because of the likelihood of double stamp duty, ie a charge would arise on each transaction. After consultations, the

Finance Act 2003 which introduced stamp duty land tax (SDLT) as a major reform of the tax, provided that certain transactions by financial institutions (each involving the purchase of a property, a lease being granted to an individual and, then a sale to the individual) are subject to SDLT once.

Ombudsman

The Financial Services Ombudsman provides an avenue for determining complaints not settled by direct recourse to the company giving rise to the complaint. Box 34.1 gives an approach to making a complaint (it is necessarily schematic and may not suit all circumstances). (The office of the FSO covers the former ombudsmen for the banks, investment institutions and the building societies.)

Part 8

Management of Property Taxation

Management of Property Taxation

26

Aim

To explain how taxation policy comes about and is implemented in to law

Objectives

- **to describe the main participants in taxation and what they do**
- **to outline the history of property taxation**
- **to explain the way in which the taxes become law**
- **to distinguish tax planning and evasion and avoidance**
- **to explain how local taxation is formulated**
- **to consider the future of property taxation**

Introduction

The overall direction and management of taxation is entrusted to the Chancellor of the Exchequer who leads HM Treasury. The day-to-day management of the system of taxation lies with the Boards of Inland Revenue and of Customs and Excise. (They are to be merged into a unified governmental body in the foreseeable future.)

As far as the property industry is concerned, local offices deal with taxpayers on day-to-day taxation matters, such as a landlord's income or corporation tax return, proposals to re-assess a business property's rateable value or a query on a business's VAT return.

The purpose of this chapter is to relate the taxation situation of a property owner or property occupier to the government structure in the UK. The next four chapters cover particular taxes and valuations for taxation.

Roles and activities

Two groups

Taxation attracts two main groups of professionals, those in national and local government and those in the private sector. In a sense they are on opposing sides both figuratively and literally. Figuratively, in that government wants the tax take and the private sector does not want to pay it. Literally, in that negotiations and disputes require the taking of sides. In general terms Box 26.1 sets out the range of roles and activities in taxation.

Box 26.1 Roles and activities in taxation

Treasury	• formulates policy • prepares draft legislation and guides it through parliament
Board of Inland Revenue	• administers taxation — assesses, collects and enforces national taxes • assesses the national non-domestic rate multiplier and receives the business rates collected by charging authorities
Valuation Office Agency Assessors (Scotland)	• assesses property for council tax valuation list • assesses non-domestic property for the rating list • appears for the official side on appeals
Office of the Customs and Excise	• deal with value added tax and other duties • register businesses for VAT • collect and remit VAT etc
Accountant	• advises clients on the effect of tax legislation • advises clients on planning for tax avoidance
Lawyer	• prepares agreements and other documentation on partnerships, companies, trusts and other legal persona • advises on disputes and conducts cases on behalf of clients • advises on probate and taxation
Valuation surveyor	• advises on values for tax purposes
Estates surveyor Building surveyor Quantity surveyor Facilities manager Estate agent	• advises on tax on disposals and acquisitions • advises on taxation in property management or facilities management • advises on taxation matters in works to property, eg on capital allowances and on programming works
Rating surveyor	• advises on assessments and appeals on Revaluation and at other times
Local government officers	• advise council on the quantum of tax required to fulfil duties and policies on discretionary policies • administer council tax — collection and enforcement
Valuation Tribunal	• hears appeals on property assessments
Lands Tribunal	• hears appeals from the valuation tribunal • takes appeals direct from the parties

---

Categories of taxes

Taxes may be categorised in different ways. In this volume the main national property taxes are identified in four groups, namely:

1. capital taxes — capital gains tax (CGT) by the Taxation of Capital Gains Act 1992 and inheritance tax (IHT) by the Inheritance Tax Act 1984
2. income taxes — income tax and corporation tax by the Income and Corporation Taxes Act 1988
3. consumption taxes — insurance tax, value added tax (VAT) by the Value Added Tax Act 1994 and stamp duty land tax (SDLT)
4. green or environmental taxes — aggregates tax, carbon emissions tax, congestion charge, workplace parking levy, and landfill tax.

Where the Act is not specified one of various Finance Acts provides for the tax.

Historical politics of taxation

Ideas, theories and practices for the taxation of property have an ancient lineage. The principal themes for national taxes in the UK have been income taxation, death duties, capital gains taxes, stamp duties, development land taxation and finally environmental taxation. They were introduced and modified (or dropped) over the centuries as shown in Box 26.2.

Box 26.2 Centuries of taxation

Income taxation	• since the beginning of the 19th century income taxation has developed	• income tax (1806 to date) • corporation tax (1965 to date)
Death duties	• since the end of the 19th century death duties have become established	• estate duty (1896 to 1965) • capital transfer tax (1965 to 1984) • inheritance tax (1984 onwards)
Capital gains tax	• Since the 1960s the taxation of capital gains has developed	• short term capital gains tax (1960 to 1965) • long term capital gains tax (CGT) (1965 to date)
Development land taxation	• since the mid 20th century taxes on development land, come and gone,	• development charge (1948 to 1950) • betterment levy (1967 to 1971) • development land tax (1976 to 1981).
Environmental taxation	• in more recent times a discernable trend has gathered pace in environmental taxation	• aggregates tax • landfill tax • energy (value added tax) • climate change levy

---

Government management of taxation

Operational management

The day-to-day management of on-going taxation is in the hands of the Board of Inland Revenue and the Board of Excise and Customs. (Box 26.1 shows the main functions and activities of taxation officials.)

Historically, the two boards each had its own body of officials but amalgamations of government functions have been introduced and an integrated department is being set up at the time of writing (November 2004).

Policy management

Essentially, principal changes for the redistribution of wealth and income in society are effected by taxation as a result of political determination and decisions. The principal government department for the generation of property taxation policy is the Treasury (headed by the Chancellor of the Exchequer). It may, however, be influenced by comment or representation from many sources, in particular, the agreed EU taxes as reflected in directives.

Proposals for change

The ways in which policy is developed into legislation is dealt with in Chapter 31. Unless a new tax is proposed, most tax changes are announced or indicated in the Chancellor's November pre-budget statement. Intense political and professional lobbying follows the November statement, particularly by the bodies representing property owners or occupiers and by the professional bodies involved with property. The March Budget and its Bill start the parliamentary process.

Annual cycle for taxation

Year by year, on-going taxation changes are effected by a somewhat regular cycle of events culminating in the year's Finance Bill receiving Royal Assent to become an Act. (In the past some years have had more than one Bill.) From time to time, other enactments contain provisions for taxation.

Tax planning

The property owner cannot do much with property without some aspect of taxation arising. Awareness of the broad nature of the way in which taxes operate is important as a basis for ensuring that the right amount of tax is paid, taking into account any exemptions, reliefs or concessions which may be available.

Taxation is complex — those in business, or in property dealing or investment would normally require the services of an accountant, and possibly a surveyor or lawyer. However, the property owner with an insight to taxation and the ability to ask the right kind of questions will be rewarded with pertinent advice and, hopefully, a lower tax bill. Box 26.3 shows the main aspects or factors of tax which need to be considered when contemplating a property transaction or works to property.

Box 26.3 Factors in tax planning

Status of the taxpayer(s)	• status affects rates of tax • availability of exemptions, etc • ensure the appropriate legal entity or entities are set up
Objectives in owning property	• objectives affect the form of taxation likely to affect the taxpayer • ensure that the objective(s) are not muddled — if so a rearrangement of status may be needed
Location of the property	• location may affect the available exemptions, etc • level of council tax may differ over local authority boundary
Type of transaction	• type of tax is affected by the transaction
Date(s) of the event(s)	• timing of events in planning for tax is important • payments of tax on time avoid interest and penalties
Type of works to the property	• need to allocate cost of works to capital expenditure or annual expenditure
Exemptions, reliefs and concessions which may be available	• exemptions, etc are creatures of other factors • different approaches to a proposal may affect availability
Options or choices of action	• alternative options or actions should be appraised on a net-of-tax basis • best financial outcome may not be the best for the client
Availability of information and assumptions for computations	• appraisals and analyses require data and assumptions • ensure that they are accurate and realistic (undertake risk of change analyses to measure variations in assumptions)
Form of report required by the client	• ensure confirmation of instructions and the date and form of the report

Evasion and avoidance

The person who fraudulently seeks to escape the payment of tax is said to be evading it and commits an offence. The authorities will pursue such a person with the view to recovering the tax and penalty — they may seek conviction in serious situations.

A person may pursue avoidance of tax with impunity, but not without danger of a dispute arising with the Inland Revenue. Depending on the nature of the property or other factors, there are several natural ways in which the legislation encourages tax planning by the taxpayer. However, the legislation also provides the Inland Revenue with ways of combating schemes or devices used to plan tax avoidance which it regards as artificial or non-commercial. For the property industry, s 776 of the Income and Corporation Taxes Act 1988 empowers the Inland Revenue to tax artificial transactions.

Locality

Within the UK locality is not usually an important factor in taxation. However, there are three types of specific location given in Box 26.4 where tax advantages accrue.

Box 26.4 Taxation advantages at specific kinds of location

Disadvantaged areas	• specified disadvantaged areas • the burden of stamp duty land tax is reduced is reduced for property buyers
Enterprise Zones (EZs)	• EZ's life expires after 10 years from designation • while EZs remain extant, 100% rating relief is available • relatively generous capital allowances are available
Freeport	• within the boundary of a freeport, value added tax is not imposed on imported goods which are processed and are then exported

---

Politics of local taxation

Much of the money used by local government in supplying goods and services comes from central government in the form of grants. Issues on the allocation of centrally held funds to the regions, counties and other tiers of local government hinge on such matters as:

*   the appropriateness of the formulae used to allocate funds to local populations
*   accountability for funds which are not derived locally
*   the burden on local government for items considered as national.

Although these matters are not directly relevant to the property industry they may have implications for it, particularly as locally raised funds are usually a tax on business and residential property occupation.

Search for a tax base

It seems that society finds it difficult to settle on the tax base for raising funds for goods and services provided locally. For nearly 400 years, the principal source of locally raised funding had been rates; largely based on the rate poundage multiplied by the net rental value of property. Thus, the current sources are:

*   rates payable by businesses (national non-domestic rate multiplier times the net rental value of the business's property)
*   council tax payable by occupiers of residential accommodation (a tax charged according to the capital value band in which the property was designated).

Various official studies into local taxation have been made and at the time of writing (November 2004) another is in progress.

Generally, alternatives to property rates have:

* if adopted, been abandoned, eg community charge
* if mooted, not adopted, eg local income tax or local sales tax.

It seems that the current enquiry has two possibilities, namely:

* widen the tax base with other forms of taxation
* extend the bands of council tax.

Future

Taxation of development value ... again?

In March 2004, a debate on a return of the taxation of development value was initiated by the Barker report on housing supply. The suggested prospect is that windfall profits or gains be charged a planning gain supplement when planning permission is granted. Box 31.1 shows that some form of betterment taxation or taxation of development value was attempted in 1948, 1962, 1967, 1974 and 1976. In a round about kind of way, development value is crudely taxed by imposts of:

* capital gains tax — when a gain is realised
* inheritance tax — when an individual dies and the land passes on (generally, development value is not exempted or relieved)
* value added tax — certain rents and sales are charged to VAT
* income taxation — on rents or premiums or both which, conceivably reflect development value
* rates or council tax — when imposed on developed land, the taxes are a charge on the total value (including any development value embedded).

It follows that a planning gain surcharge may need to allow for double taxation relief. However, an alternative approach might be site value rating (but that is another story).

National Taxes and Property

27

Aim

To demonstrate the impact of national taxes on property ownership and on transactions involving property

Objectives

- to briefly consider the historic and present day scope for taxation
- to review the impact of each of the principal taxes on property owners and their activities
- to explain the effect of the principal exemptions and reliefs of each tax on property ownership

Introduction

All taxation is statute-based and the national taxes are those which are imposed by central government for national purposes: traditionally local taxes are imposed on the population of a local area by the local authorities and other local bodies. In this chapter the principal taxes covered are:

- income tax and corporation tax
- capital gains tax
- inheritance tax
- value added tax
- stamp duty land tax (SDLT)
- various environmental taxes.

Although aspects of what is now called environmental taxation have existed for many years, the topic has gained in stature in fairly recent times as a response to a number of concerns, namely:

- the contaminative impact on the global environment of traditional energy policies and usage
- the dependency on oil and coal as non-sustainable or renewable sources

- the developing geo-political threats to traditional energy supplies
- the consumerism and its impact on non-sustainable resources.

As a result property interests have been affected in with new taxes which, in general, are designed to alter perceptions and practices number of ways, including:

- traffic congestion taxes
- taxes to change the travel to work balance
- aggregates levy to encourage recycling
- landfill taxes to encourage better waste management, particularly recycling
- taxes to encourage investment in energy management and in renewable energies.

Roles and activities

Box 27.1 sets those involved in advising on national taxation matters.

Box 27.1 Professionals and others involved in national taxation

Taxpayer	• runs a business, deals in land, invests in land or owns a house • enters into transactions in land or carries out work to property
Accountant	• advises on taxation matters • advises on tax planning
Lawyer	• advises on law of taxation • advises on probate
Valuation surveyor	• prepares valuations and appraisals for taxation purposes • advises on the application of exemptions and reliefs
Quantity surveyor	• prepares appraisals for capital allowances • advises on costs and taxation
General commissioners	• hear appeals against income taxation and other assessments
Special commissioners	• hear appeals on complex taxation matters
Lands Tribunal members	• determine appeals on valuation matters for taxation
Inspector of Tax	• makes assessments for income taxation and capital gains tax
Collector of Tax	• collects taxes and enforces payment

Income taxation

Property as part of the tax base

Income tax and corporation tax is assessed according to the amended provisions of the Income and Corporation Taxes Act 1988. Schedules (some with cases) in the body of the Act give the computations.

The property industry generates considerable income tax and corporation tax, including profits and gains from:

- rents and premiums under leases (schedule A)
- annual interest on loans for the purchase or improvement of property
- annual payments for wayleaves and easements
- profit derived from transactions caught under statutory provisions for anti-avoidance schedule D case IV (see s 776)
- dividends from shares held in property investment companies, property dealing companies and contracting companies (schedule F)
- salaries, wages and pensions (schedule E).

Individuals, business partners and certain other persons are taxed to income tax on their net incomes (but not capital gains) from various property sources.

Taxation of companies: corporation tax

Companies and certain other organisations are taxed to corporation tax on their profits from various property sources of income and on gains. The same schedules are used for a company's incomes as for individuals and others but the rates of tax are different as are ways of payment and the dates when it must be paid.

Capital allowances

Procurement of new buildings and improvements to existing property is capital expenditure which is not normally allowable against annual profits. However, there is a statutory regime for the depreciation of specified assets.

Owners and managers of qualifying buildings and other assets should ensure that professional advisers or staff categorise expenditure on buildings so as to identify what is allowable and make the claim at the appropriate time.

Thus, notionally the fabric of a building wears out during the life of the building; similarly for plant and machinery, ie they depreciate. The Capital Allowances Act 2002 provides that for each year, part of the capital cost of certain buildings, plant and machinery and other assets may be treated as an expense of running the business — so accounting for the depreciation.

Writing down allowances

The Act gives various rates of depreciation in the form of writing down allowances. The statutory depreciation is different from the accounting depreciation allowed in the financial statements of a business. Basically, the capital cost of an asset has either reducing balance or straight line depreciation. The first is a deduction calculated as a percentage of what was left after deduction in the previous year. The second is a fixed percentage of the original cost.

Box 27.2 Capital allowances available to the property industry under the Capital Allowances Act 2002

Part 4	Agricultural buildings
Part 9	Dredging
Part 4A	Flats above shops and other commercial buildings
Part 10	Dwellings on assured tenancies
Part 4	Forestry buildings
Part 5	Mineral extraction

Capital gains tax (CGT)

Capital growth of property and CGT

Capital gains from the disposal of the whole or part of a property may give rise to CGT under the Taxation of Chargeable Gains Act 1992. However, capital losses of capital may be allowable against such gains. Disposal covers several events, including:

- the sale or exchange of a freehold
- the grant of a lease at a premium (income taxation may also arise)
- the sale of a lease
- the gift of property (within seven years of the gift an inheritance tax liability may arise)
- the proceeds of a claim under an insurance policy.

Basic calculation

For a property disposed of recently, the basic calculation for CGT is as follows:

		£
	Consideration or Market Value	
Less	Incidental costs of disposal	
	Cost of acquisition or market value	
	Incidental costs of acquisition	
	Enhancement expenditure	
	Costs of defending title	_____
Hence:	Basic Gain or Loss =	_____

Market value is used where there was a gift or some other disposal not at full value. Also, it is used for rebasing, ie all gains are now deemed to have accumulated since 1982. As a result the open market value at 31 March 1982 is used instead of the appropriate figure for those acquisitions which were at an earlier date.

The basic calculation is supplemented for allowances of indexation (part of the time) and for taper relief (part of the time). Of course if the taxpayer is eligible to roll-over relief (or retirement relief for disposals before 6 April 2003) it may be claimed.

Exemptions and reliefs

The CGT regime provides a number of exemptions and reliefs, including:

- gifts to one's spouse
- deemed disposals on the death of an individual
- a individuals sole or main residence including grounds of up to 0.5 hectare
- taper relief which reduces liability according to when the asset was acquired.

Taper relief replaced indexation but the latter may also be available for acquisitions prior to the change.

Payment

A payment is the CGT on the net capital gain in the year, ie capital gains less any allowable losses, including any brought forward from previous years. The payments must be made by the due date.

Inheritance tax (IHT)

Nature of IHT

IHT is mainly the concern of individuals, although trustees will be concerned about the impact a death may have on the trust's assets should, for instance, the grantor dies. Also, a family business may be affected should the proprietor or a principal shareholder die.

Broadly, IHT is a potential gifts tax and a death duty but it only becomes due on the death of an individual. The value of the deceased's estate at the date of death is ascertained together with the value of any gifts made in the seven years prior to death. Tax is levied on the net value of the estate.

Net value of the estate

The tax is charged on the net value of the estate which is the gross value of the assets after deducting the value of the items. Items or allowances left out of account include:

- gifts to the surviving spouse
- gifts to heritage bodies
- gifts to charities or for the benefit of the public
- gifts made more than seven years before the date of death
- reasonable funeral expenses.

Lifetime and death transfers

A gift of property made during lifetime is known as a potentially exempt transfer, becoming exempt if the giver does not die within seven years of the date of the gift. If the donor dies within the seven years the gift is treated as part of the estate passing at the date of death. Where the deceased had survived the gift for a year or more relief is given (see below).

Net estate

Personal representatives need ascertain the net value of the estate which is the aggregate value of all assets which are not exempt. Also, certain items are allowed or deducted from the value of the positive assets, such as mortgages.

The net value of the estate is:

The aggregate of such non-exempt assets as:
- freehold and leasehold property
- chattels
- intangible property, eg shares in a property company
- loans payable to the deceased.

Less: the aggregate of such liabilities and other amounts as:
- mortgages and loans payable out of the estate
- income and capital gains taxes due to be paid
- reasonable funeral expenses.

Exemptions and reliefs

There are numerous exemptions and reliefs from IHT but many of them are not directly related to the property industry. However, some concern favoured assets, eg agricultural land and buildings. The use of exemptions and reliefs in tax planning is an important way of protecting an estate against the IHT (and perhaps other taxes). Most of the exemptions and reliefs concern individual taxpayers but in some circumstances a trust or business may be affected — these matters are dealt with in the context of other taxes under the section on tax planning below.

Payment

Payment of IHT must be made by the due date but arrangements may be made to phase payments over 10 years (paying interest on outstanding tax). Alternatively, important heritage property may be given in lieu of cash on a special appraised basis.

Value Added Tax (VAT)

VAT was introduced in 1973 when the UK joined the European Union. It is now consolidated in the VAT Act 1994. The tax arises when one of about 1.6m registered businesses makes a taxable supply, including many property transactions or when works to property are undertaken. An outline of the principles is given in this section together. Box 27.3 gives examples of property transactions or works to property where the tax is likely to arise (when the supplier is registered). However, some instances enjoy a reduced rate, ie 5% rather than the 17.5% standard rate, eg works to property in disadvantaged areas.

Box 27.3 VAT liability on property transactions and other supplies

Professional services	• are standard rated, including those of architects, lawyers, planners, property managers and surveyors
Construction works	• demolition services are standard rated • new build is standard rated • maintenance and repairs are standard rated • refurbishment is standard rated
Building materials	• subject to VAT at the standard rate
House and flats — new build	• such new build is an exempt supply
Car parking space	• parking charges are subject to VAT at the standard rate
Landlord's option to tax	• rents and service charges are subject to VAT at the standard rate once the landlord opts — applies to all leases of the estate
Service charges	• subject to VAT at the standard rate

Payment by the last consumer

The initial incidence of VAT is borne by the customer in each transaction. If registered the customer who pays VAT on goods supplied and then supplies the goods to the next customer charges the tax - ultimately the last (customer) consumer bears the incidence of VAT. Thus, VAT is an *ad valorem* tax payable by the last customer in the supply chain to a registered supplier. In principle, the supplier recovers any tax paid previously (on supplies received) from the tax received from the customer; remitting the net amount to the VAT office. (If more tax was paid than was received, a claim for the difference is made to the VAT Office.)

Of course, an individual transaction is not treated individually — a registered supplier's transactions are aggregated each quarter and the net amount paid (or received) accordingly.

Input tax and output tax

Each supplier invoices a customer and the invoices must show the amount of VAT payable (output tax). Similarly, when the supplier receives goods or services from another supplier the invoice will show the amount of VAT to be paid (input tax).

Exempt supplies

A registered supplier may make a supply of goods or services which is an exempt supply, ie no tax is charged to the customer; but the supplier cannot claim VAT paid previously on goods and services.

Rates of tax

Rates of VAT are either standard (17.5%), special (5%) or zero (0%). Where a supply is charged or rated at one of these rates the supplier may recover input tax. (A zero-rated supply is in contrast to an exempt supply.)

Stamp duty land tax (SDLT)

When and individual or organisation buys, leases or enters into certain other kinds of property transactions a liability to SDLT arises on that person. The buyer, etc is required to serve a land transaction return and to pay the tax in the allotted period, generally 30 days. One effect of not paying is that the land involved cannot be registered at the Land Registry.

Stamp duties have been in existence since the Stamp Act 1891. However, the property form of the tax was changed when SDLT was introduced by the Finance Act 2003. One change was to make the tax a charge on transactions rather than the documents giving effect to the transactions. The range of rates is the same for residential and business property but, whereas the threshold for 1% is £60,000 for the former, the threshold for business property is £150,000. The two scales go to £500,000 and above, when it is 4%. The buyer must inform the tax authority

Exemptions and reliefs

Exemptions and reliefs are available for certain transactions, eg purchases in disadvantaged areas. The effect of reliefs is, generally, to reduce the amount of tax in different ways according to the type of transaction (see Box 27.4) for the persons who enjoy exemption or a relief. It is useful to note that each has conditions which must be observed.

Environmental taxation

European Union taxation policies

Apart from taxation which had to be incorporated into UK law when the UK joined the EU, the UK has resisted the development of EU taxation policy. Nevertheless a number of EU-driven taxes have been developed and are now law in the UK.

The taxation policies which have been incorporated into UK law:

- VAT
- Customs and Excise duties
- landfill tax.

Box 27.4 Stamp duty land tax — some exemptions and reliefs

Disadvantaged areas	• certain transactions to encourage investment in disadvantaged areas are given relief
Exchange with alone seller	• where a house-builder exchanges a house with a lone owner occupier • conditions as to the garden's area • consideration is treated as nil
Company Employee	• company buys an employees house when the latter relocates — exempt • when employee relocates (at least for a year) — exempt
Compulsory purchase	• acquisitions by a public body are exempt
Planning obligations	• transaction under a planning obligation are exempt for a public body
Crofting community	• a special calculation applies where a crofting community buys the superior property
Registered social landlords	• transactions by RSLs are exempt
Right to buy • flat owners	• covers the transactions under the Landlord and Tenant Act 1987 and the Leasehold Reform, Housing and Development Act 1993
• commonholders	• a special calculation is provided
Charities National purposes bodies	• transactions are exempt
Organisational restructuring transactions	• exemptions for de-mutualising insurance companies, building societies • restructuring etc banks, building societies and groups
Limited liability partnerships	• transfers on incorporation are exempt

Aggregates levy

The mining and minerals industry supplies the construction industry with the newly exploited aggregates, such as gravel, rock and sand, used in the construction of buildings and structures. In addition, part of the output of aggregates comes from the recycling of hardcore and other waste. In an attempt to lessen the environmental impact of the extraction of aggregates, the government introduced a tax on the purchase of newly extracted aggregates — hopefully making recycled aggregates relatively cheaper. The tax is charged by the tonne of newly exploited aggregates whether the source is land-based, offshore or imported.

Landfill taxation

Problems with the traditional way of dealing with waste have arisen where household waste and that from industrial and other sources have been buried in landfill sites. The sites are gradually becoming full and the environmental damage of uncontrolled landfill has become more widely recognised. As a result alternative ways of dealing with waste have been introduced, eg incineration or recycling. Also, the Finance Act 1996 introduced landfill tax under which the levy is made at a low rate per tonne for inert waste but much more severe for active waste. As a result, recycling has become much more common.

Renewable options

An artificial or false market has been created by the government for producers of renewable energy. In the past, the suppliers of energy to homes, industry and other buildings have had to buy most of their electricity from mainly non-renewable sources. Now they are required to buy a portion from the producers of renewable-sourced energy. Occupiers of property who produce their own electricity from a renewable source will find that they may be able to sell any surplus on the artificial market.

Taxation of carbon

In certain industries, companies have been preparing for another artificial or false market — that in carbon emissions. Those who burn oil, gas and other organic fuels produce carbon and other contaminants when creating energy for heating, lighting and powering homes and other buildings. The carbon generated in this way is thought to contribute to global warming and the loss of the ozone layer. For this reason, the government introduced the climate change levy whereby an industry has an aggregate of carbon emissions allocated to it (from the estimated national total).

Each separate business in an industry is allocated a portion of the emissions from the industry total. During the course of business some companies will not use all their emissions while others will need to use more than their allocation. The companies may trade in emission vouchers but some companies will be induced to invest in plant and machinery to become more energy efficient and so make les emissions — any surplus emissions may be sold on the emissions market.

Workplace parking levy

The workplace parking levy is designed to charge a tax on parking spaces at workplaces. Its intended effect is to reduce motor travel and hence congestion. The tax was provided by the Transport Act 2000 on a discretionary basis, ie local. Authorities may choose (or not) to implement it.

Congestion charging

As a result of the London Assembly introducing the congestion charge for a zone in London, it is claimed that congestion is down by about 30% and that millons of pounds have been raised to fund transport improvements in the capital.

Tax planning

Property owners and occupiers, generally, have many opportunities to reduce the amount of property tax they pay by careful planning. The scope for tax avoidance (not to mean tax evasion which is, of course, illegal) is one which will usually require the advice of one or more of an accountant, lawyer, valuer or quantity surveyor.

Most transactions involving property — buying, selling, building, improvement, repair or maintenance — proffer opportunities to consider whether tax planning is appropriate. Sometimes the choice of location for a new build or a purchase may be important.

Box 26.3 briefly indicates some of the ways in which tax may be taken into account in property transactions. No attempt has been made to marry property tax matters with other matters which may be important in tax planning. Thus, some aspects of tax planning touch upon other matters which are dealt with elsewhere in this volume, ie tax planning is alluded to in a number of chapters. However, it cannot be over-emphasised that the subject requires specialist professional advice.

The main topics dealt with elsewhere concern:

- the status of the tax payer
- the estate or interest which is owned
- the nature of the business (or objective of ownership)
- the type of building
- the location of the building (see Box 26.4)
- the nature of the works
- the availability of tax exemptions, reliefs and concessions.

The last topic is, in effect, the basis for the others but any proposed transaction or works may have tax implications under one or more of topics. Before embarking on a property acquisition, development or other transaction, particularly where there are choices of alternative routes and, hence, outcomes, the tax implications should be investigated to ascertain the best net-of-tax result. Of course that may not be the choice but the taxpayer should know the opportunity cost of the action taken.

Status

Exemption from tax has been noted elsewhere for various kinds of legal personality. When an individual sets up a company, a trust or some other entity, professional advisers should examine and explain the taxation implications.

Estate or interest owned

The estate or interest owned may give a measure of tax saving but this is not a common feature of property transactions. Instances include:

- a lease granted at a premium; income taxation does not arise when the term is greater than 21 years
- the nature of the interest may be important for capital allowances
- similarly, for SDLT.

In all of these situations it is important to consider the taxation implications before any contract on the transaction. A professional adviser should be able to give advice on the best approach.

Nature of the business or objective

From time to time the government, wishing to encourage investment in a type of property, will make capital allowances or other exemptions or reliefs available. Similarly, plant and machinery or other features of a building may enjoy relief, eg thermal insulation.

Type of asset or favoured asset

Many assets enjoy exemption and relief under various taxes. For instance, the avoidance of IHT is an important concern of individuals who are property owners. One way of achieving some avoidance is to own a favoured asset; a term denoting any property which is either exempt from IHT or enjoys relief from the tax. In Box 27.5 the concept of favoured asset is expanded to include other taxes giving exemption or relief. The box merely indicates the existence of the exemption or relief — each exemption must satisfy conditions and limitations; so professional advice will usually be required to ensure the effectiveness of any tax planning.

Box 27.5 Favoured assets for inheritance tax, particularly, but also other taxes

Owner occupier home	• an individual's sole or main residence is exempt form CGT unless it has a garden of more than 0.5 hectare • rent-a-room income tax relief is available to the owner who rents out a room with certain limits on rent
Agricultural land and buildings	• IHT — agricultural property relief is afforded to the agricultural property, valued assuming a perpetual covenant • for tenanted farms, the landlord may obtain relief in a similar manner • capital allowances are given on agricultural buildings
Woodlands	• IHT — exemption afforded while the living trees are owned (possibly through generations), and not harvested • Income tax — exemption from schedule A on forestry woodlands (also, for land being prepared for forestry) • IHT — short rotation coppice enjoys agricultural property relief • CGT — timber is taken out of the computation for CGT, ie it is exempt • capital allowances are enjoyed on forestry buildings
Heritage property	• IHT — exemption may be enjoyed on heritage property — so specified by the Treasury
Heritage objects	• heritage objets d'art may enjoy exemption when held with a heritage property — so specified by the Treasury • land of some beauty and so may qualify
Business assets	• exemption is afforded to some business assets
AIM listed shares	• certain AIM shares are given exemption from IHT

Nature of the works

A portion of any capital works to certain buildings and structures which are owned by business occupiers or property investors may enjoy capital allowances. Not all capital works are covered but attention to design and choice of materials or functional arrangement may make a difference. The nature of the works and the need to highlight the entity paying for them are important features of tax planning for capital allowances.

Generally, works which have a functional contribution to the business or enhance the ambience of the setting where business takes place will be allowable. However, works which are the setting or part of it will not be allowable. When the works are a mixture of repair, maintenance and capital works, careful planning is needed to record and apportion the expenditure.

An estate's buildings may be let on different kinds of leases so that repairs and maintenance to one building (which cannot be allowed for income tax purposes on the income from that building) may be allowable against the income of another property. These planning arrangements may require an analysis of:

- the types of lease each property is let on
- the proposed timing of works to each property
- the income from each property and hence any deficit due to the works.

Council Tax and Residential Property

28

Aim

To describe how local taxation affects the occupiers and owners of domestic property

Objectives

- to explore the nature of local taxation as it affects domestic property
- to identify the roles and activities in rating
- to describe the billing authority's functions
- to examine how the exemptions and reliefs apply

Introduction

Before the introduction of council tax from 1 April 1993, community charge was imposed on occupiers of dwelling; the community charge had earlier replaced domestic rates. The Local Government Finance Act 1992 established council tax as an annual charge on owner-occupiers or tenants of residential property. It is payable on 1 April, or by instalments, and is subject to many exemptions and reliefs, including means tested reductions. It applies in Great Britain. Northern Ireland has rates.

Apart from occupiers of chargeable dwellings certain other persons may be liable, including:

- the personal representative of a deceased occupier
- the owner of a vacant property
- the owner of a composite property
- certain other owners who are designated as liable instead of the occupiers.

Roles and activities

Box 28.1 describes roles and activities of council tax players in the public and private sectors.

Box 28.1 Roles and activities in council tax

Billing authority	• determine the amount overall amount to be collected including imposed precepts
	• determines individual liabilities
	• charge, collect and enforcement of council tax
	• use discretion to propose changes to a hereditament's inclusion in a band
Precepting authority	• determine the amount of its precept
Valuation Officer of the Valuation Office Agency (VOA) Assessor of Regional Council in Scotland	• value properties for the revaluation
	• prepare the valuation list and publish it
	• receive proposals to change bandings
	• value altered properties for a change of banding
	• attend appeals and present the official case
Rating surveyor	• advises owners and occupiers on the value of their property
	• if instructed, value and negotiate with the VO or the Assessor
	• if instructed, present evidence at any appeal
Magistrate	• hears enforcement cases
	• issues liability orders
	• makes committal orders
	• hears appeals against distress
Bailiff	• following the making of a magistrates liability order, may execute a distress procedure
Valuation Tribunal	• hears appeals
High Court	• hears appeals from the Valuation Tribunal
	• hears important cases directly from the parties
	• appeals go to the Court of Appeal

Council tax and dwellings

Houses, flats, caravans and houseboats used as dwellings are covered by council tax. Also, the part of any business property which has some residential accommodation, eg a caretaker's flat, will be caught for council tax as composite property. (The approach to valuing composite property is briefly discussed in Chapter 30.)

Where a garden, outhouse or other appurtenances is with a house it is included to the extent that it may enhance the value of the house, even one on the other side of a road. Generally, however, the right to use the garden is not regarded as being taxed as such. An exception arises where the garden or a part of it is the pitch for a caravan used as a sole or main residence. Also, a separate garden some distance away would not be included with the house (and would be outside of the council tax regime).

Other dwellings

Certain other residential accommodation is caught under the 1992 Act, including:

- nursing homes
- hotel accommodation for resident guests
- a farmhouse and workers' cottages (although agricultural land and buildings are exempt from business rates).

Holiday cottages and caravans and the like are normally subject to business rating.

Prescribed classes of dwellings

Certain properties are by order a prescribed class of dwelling and the owner is liable for council tax. The order covers:

- residential care homes
- religious communities
- houses in multiple occupation
- resident staff
- ministers of religion
- houses occupied solely by asylum seekers.

Administration

Billing authorities

The billing authority determines the unit amount of council tax having had regard to:

- its estimated spending on statutory duties and discretionary opportunities
- government grants and the like
- the total number of adult residents
- the number of adult residents excluded from the total
- amount of any reduction caused by means tested council tax benefit
- the amounts sought by the precepting authorities
- estimates of other receipts and obligations, eg interest on loans.

The billing authorities are district councils, unitary authorities, metropolitan district councils and London borough councils. They charge, collect and enforce payment of the council tax from occupiers of dwellings and others.

Precepting authorities

Certain other authorities, eg county councils, police authorities, town councils and parish councils inform the billing authority of their council tax requirement and receive a share of the council tax as a precept.

Enforcement

When a taxpayer fails to pay the billing authority may enforce payment by one of the following:

- by obtaining a liability order from the magistrates court
- by petitioning for the defaulter's bankruptcy
- by applying to the magistrate for committal to prison (used when insufficient money is raised by the bailiff.

In cases where the billing authority has obtained a liability order it may proceed by one of the following:

- use distress procedures
- impose an attachment to earnings or deduction from a job seeker's allowance or income support
- apply to the county court for a charging order to register a charge against the defaulter's property.

Council tax payers

Occupiers and landlords

Owner occupiers or tenants of property are liable to pay council tax for it in April and October of each year. Alternatively, it may be paid by instalments, without interest, over 10 months from April.

Occasionally, a landlord will be liable for the tax. Cases include:

- the lease provides that the rent is inclusive
- the property is vacant and it has become liable
- the property is in multiple occupation.

In the first situation, in the event of non-payment, the liability remains with the occupying leaseholder who, as such, cannot contract out of the statutory liability to pay the tax.

Business property — residential occupiers

Where a factory or shop, for instance, contains residential accommodation for staff a liability an assessment would have been made and liability arises on the occupier or the employer (depending on the terms of employment).

Owners of vacant property

Completion of newly built property or the vacation of existing property will usually result in a period of six months exemption. After the six months the owner becomes liable to council tax at 50% of the full liability.

Young tenants

Where a tenant is under 18 years of age, the landlord is liable but may be eligible for a 50% relief.

Exemptions and reliefs

Box 28.2 indicates the exemptions and reliefs which are available, giving brief details of each.

Box 28.2 Council tax — taxpayers and persons who are exempt, etc

Occupier and spouse	• jointly and severally liable
Two or more persons occupy the dwelling as a household	• they are treated as if two persons occupy the property
Lone occupier	• 25% reduction in amount due
Landlord	• contractually liable to the tenant to pay the tenant's tax on behalf of the tenant, ie where the lease provides for the rent to be inclusive of council tax • property is unlet and unfurnished — 50% liability may arise after six months
Vacant property and the owner is:	• in prison • in hospital or a care home • deceased (but only for six months after the grant of probate) • a student • a member of the clergy
Owner of vacant property	• where occupation is not allowed by law
Mortgagee of vacant property	• the property has be repossessed
Person who is physically disabled	• a relief may be available
Unoccupied second home	• a 50% relief is afforded (but see "Future" below)

Council tax calculations

Unlike business rates, which are based on a single national uniform non-domestic rate, the level of council tax is fixed by the billing authority. The amount paid for a dwelling depends on its band ie one of eight bands (band A to band H.). The assessed capital values for each property fall into its band of capital values (see Box 28.3 for the bands in both England). Occupiers of property in a band are billed council tax according to its band's allocated proportion of the band A amount.

Box 28.3 Council tax bands and the proportions on band A for each band

Band	Value as at antecedent date				Amount payable (as multiple of the amount payable for a band A property)
A	Up to	40,000			1.00
B	from	40,001	to	52,000	1.27
C	from	52,001	to	68,000	1.33
D	from	68,001	to	88,000	1.50
E	from	80,001	to	120,000	1.83
F	from	120,001	to	160,000	2.16
G	from	160,001	to	320,000	2.50
H	over	320,000			3.00

Government capping

Although the billing authority fixes the amount of tax to be collected, the government is empowered to restrict an authority's council tax take by capping it.

Council tax assessments

Valuation to capital value

Each dwelling is assessed to capital value. The Valuation Officer (or Assessor in Scotland) is responsible for the every assessment, subject to appeal to the valuation tribunal (see Chapter 34).

Appeals

Appeals are made to the valuation tribunal on valuation matters. On a national revaluation the right arises after list has been published by the VOA and must have been made within six months of 1 April 1993. Subsequently opportunities can arise in particular circumstances, such as when there is a material reduction in the capital value or when the property has been purchased.

Generally, the matters giving rise to an appeal are:

- the banding is inappropriate
- the property is not a dwelling
- the property has been altered

- the property has been demolished
- an altered banding is not acceptable.

Inappropriate banding

On a revaluation the new list is published by 31 December to come into operation on the 1 April. An owner-occupier, landlord, tenant (as a taxpayer) or billing authority then has six months from 1 April to 30 November to appeal against the banding and seek a change. Appeals against the 1993 Revaluation can, of course, no longer be made.

Property is not domestic

Where a property is a business property or otherwise not domestic it should not be included in the rating list.

Alterations to the property

Physical alterations to a property, eg partial demolition, may reduce its value below the lower threshold of the band in which it was originally placed. If so, an appeal should result in a placement in a lower band.

Demolished property

Demolition of a property will result in its removal from banding.

Altered banding is not acceptable

If an altered banding is not acceptable the person aggrieved may seek to agree a new banding with the valuation officer or appeal to the Valuation Tribunal.

Practical pointers

New buildings, improvements and alterations have implications for taxpayers which should be dealt with speedily. Box 28.4 highlights some of the practical points which need to be addressed.

Future

The government is reviewing the future of council tax. Hints of possible changes, such as replacement with a local income tax, have historically always been around. However, until a definitive policy statement is issued by the government and the policy is implemented, council tax is likely to remain without substantial change.

Box 28.4 Council tax and other issues arising when works are carried out to property

Construction of new dwellings	• check for exemption or relief — may depend on the each occupier' status • likely need to negotiate with the Valuation Officer on the assessments • a liability arises when a chargeable dwelling becomes occupied or if unoccupied, on expiry of a period of six months after the property became substantially complete • occupation by the builder of any site office or sales office or both may be liable to business rates
Self-build by the site owner	• living on the site in a caravan, for example, likely to result in its assessment to council tax • living in the partially completed new dwelling, needs a certificate of habitation from the building control officer • similarly, habitation may result in assessment and liability for council tax
Alterations to a building	• any change in capital value may result in possible change in prospective banding • improvements (result in increase in value) only result in a possible step up of banding on the sale of the property or on national revaluations which are now likely to occur every 10 years (the government does not want to discourage improvements • "negative" improvements may result in a step down in banding — make a proposal to the VOA
Demolition of a building	• when the building is first vacated seek cessation of payments to tax • partial demolition may prevent incidence of tax under vacant property charge or at least reduce the value of the property • after demolition make a proposal to remove from the list
Repairs and maintenance	• repair and maintenance should not result in an increase in value • exceptional standards of repair and maintenance, so that the dwelling is outstanding, may increase the value over others (a possible revaluation issue but unlikely)
Adaptations for a person with a physical disability	• the adaptations to the chargeable dwelling may give an opportunity to claim a relief
Conversion of a single dwelling into flats	• may affect the banding • report to the VOA; each self-contained flat will give rise to a separate council tax liability • separate assessments for each flat may mean that the burden for the owner (retaining a flat) will be reduced
Conversion to flats of the upper floors of a shop or offices	• the flats will become chargeable dwellings • business rating (and an composite property element) will, no doubt, change • planning permission will normally be required • capital allowances may be available
Adaptations (probably minor) for rent a room income tax relief	• should not result in any change to the banding status • single person discount (if being claimed) will end when room is let (unless to a student, or person under 18 or lodger in receipt of income support or job seeker's allowance
Conversion of a dwelling to a holiday cottage etc (or a change of use)	• takes a hitherto chargeable dwelling into business rating • business comes within the income taxation regime, eg possible claim for capital allowances

Revaluation

An English national revaluation begins in 2005. Bills based on the new valuations will be issued 1 April 2007. Revaluations on a nationwide basis are now due to take place every 10 years to enable continued accuracy in relative values of each property.

Such valuations do to not in themselves give rise to an overall increase in council tax payable. However, such an increase may be experienced where the capital value of a residence or group of residences has, for any reason, increased disproportionately to other values in the billing authority's area.

In Wales the recent new revaluation comes into force on 1 April 2005 so new bills and a new cycle of appeals will begin from that date. Similarly, England's new revaluation will result in a new cycle of appeals from 1 April 2007. In Scotland a Bill is progressing through the Scottish Parliament to abolish council tax and introduce service tax (an income tax).

Transitional arrangements

The revaluation is expected to be accompanied by the introduction of transitional provisions that will phase in the impact of any significant increase or decrease in amount payable on any individual property or group of properties.

Changes to banding structure

The English revaluation is also expected to be accompanied by the addition of extra bands. This is expected to redistribute the burden of council tax and will mostly impact on residences of the highest value.

Unoccupied second homes

Local authorities have recently been given discretionary powers to partially or wholly remove the 50% discount available for unoccupied second homes.

Power to reduce council tax

Local authorities have recently been given discretionary powers to partially or wholly reduce the council tax due in any particular case or class of case as it thinks fit. For example, a few authorities have used this power to reduce the council tax burden on pensioners.

Business Property and Rates

29

Aim

To describe how local taxation affects the occupiers and owners of non-domestic property

Objectives

- to explore the nature of local taxation as it affects non-domestic property
- to identify the roles and activities in rating
- to identify the method of valuation which applies to particular property
- to examine how the exemptions and reliefs apply

Introduction

Business rates are charged on the occupation of non-domestic property, raising about £20 billion as a result of the Local Government Finance Act 1988. Essentially, the system is that which has applied to non-domestic property since the turn of the 19th century, ie with assessments to net rental values: whereas council tax is based the capital values of the dwellings (except in Northern Ireland).

Roles and activities

The roles and activities are similar to those for council tax but there are differences in detail. Box 29.1 gives the business rating context of individuals and organisations together with what each does.

Rating business property

Occupier's liability

It is important that in the first instance the occupier of a business property is liable and charged business rates on a multiple of the previously assessed rateable value (an annual rent) of the property.

Box 29.1 Roles and activities in business rating

Charging authority	• determines individual liabilities • charges, collects and enforces payment of business rates • remits collected rates to the government • uses discretion to make proposal to the VOA for a change to a hereditament's rateable value
Valuation Officer of the Valuation Office Agency (or equivalent in Scotland and Northern Ireland)	• values properties for national revaluations every five years • prepares the rating list and publishes it • receives proposals to change rateable values • attends appeals and presents the VOA's case
Rating surveyor	• advises owners and occupiers on the value of their property • if instructed, value and negotiate with the VO • if instructed, present evidence at any appeal
Magistrate	• hears enforcement cases • issues liability orders • makes committal orders • hears appeals against distress
Bailiff	• following the making of a magistrates liability order, may execute a distress procedure
Valuation Tribunal	• hears appeals on matters of fact
Lands Tribunal	• hears appeals from the Valuation Tribunal on matters of fact and of law • hears important cases directly from the parties • appeals go to the Court of Appeal and hence to the House of Lords

---

The meaning of occupier has exercised the minds of judges over many years; so much so that a concept of rateable occupier, based on the principles of rateable occupation, has evolved from the court cases.

Principles of rateable occupation

The four evolved principles hinge on four descriptive words for occupation, namely:

- actual
- exclusive
- beneficial
- permanent.

Space does not permit a detailed account of the dozens of cases which explore these words but the subject is fascinating. Usually it is obvious who is actually in occupation of a rateable property. However, where it is unclear, it is important to note that the legal terms have definitions for the four principles above which are far removed from their ordinary grammatical use; as such no attempt should be made to interpret them from their ordinary meanings.

Hereditaments

Where a property is rateably occupied in parts, ie by more than one person and each satisfies the principles of rateable occupation, there will be an hereditament for rating purposes for each person. The Local Government Finance Act 1988, s 64 considers the scope of hereditament. Each person will, therefore, be billed on a separate business property (being a part of the whole property) which will have been previously assessed. Again, a large number of cases have established the meaning of hereditament, eg *Westminster City Council* v *Southern Rail Company Ltd* (1936).

Assessments

For the national revaluations, the assessed rental value is determined by officers of the Valuation Office Agency every five years, ie at time for the revaluation of all business property in the country.

National non-domestic rating multiplier

Every year the multiple or national non-domestic rating multiplier is announced by the Treasury for use by the local charging authorities.

Rates payable

Charging authorities apply the national non-domestic rating multiplier to the rateable value of each non-domestic property in their area so fixing the rates by the occupier for the year. However, some types of property and some organisations are exempt or enjoy a reduction in rates, ie a relief is afforded (see below). It may be noted, however, that empty property may become rateable at 50% under the Non-domestic Rating (Unoccupied Property) Regulations 1989 (SI 1989 No 2261).

Collection and enforcement

The charging authority collects the rates but if an occupier fails to pay the rates the billing authority is empowered to enforce payment. The total amount of the rates is credited to the government which then allocates the funds to government departments for redistribution to local authorities in accordance with national policies for programmes, projects and on-going operations (see Chapters 31 and 32).

Business improvement districts (BIDS)

Councils may now impose an additional levy on the business rates to pay for a specific improvement that has been requested by the ratepayers themselves, such as the provision of additional local security or a CCTV service. A BID can only be created if the majority of the businesses within the BID district vote in favour of the scheme.

Revaluation for rating

Revaluation — general

Every five years a new non-domestic rating list is published by the Valuation Office Agency. Under the current system new lists were published at the end of 1989, 1994 and 1999. The date provided for publication is 31 December — three months before the list comes into force on 1 April. The rateable values for the forthcoming rating list have already been published. The current rating list came into force on 1 April 2000 and the new one is due to come into force on 1 April 2005. When the rating list comes into force the new assessments are not fully imposed immediately — a transitional period is arranged so that ratepayers do not experience excessive changes in the amount they pay (see below).

Revaluation — current

The Valuation Office Agency has prepared the rating list for 2005 and has already notified occupiers and others of the availability of the assessment in respect of their property. Many but by no means a high proportion may well be intending to investigate the assessment to see whether a case might be made for an appeal.

Programme of work

National revaluations take place every five years. The Valuation Office Agency is responsible for the revaluation of all business and other non-domestic property, taking about two years to complete the work. It may be noted that whereas a dwelling is assessed to capital values and then put into bands, a non-domestic property is valued to net rental value (but not banded). (The next chapter, Chapter 30 Valuations for Taxes, summarises the basis of valuation for non-domestic property.)

Property types and valuation methods

Rating cases probably provide the widest and deepest study of the traditional methods of valuation; they are dealt with in Chapter 24. In valuing each type of property for rating, the valuer will use a particular method of valuation, and in some instances a valuer has been known to use two or three methods. Box 29.2 gives a selection of property types, the methods of valuation used to value them.

Rateable property (hereditaments)

For the most part the property or hereditaments shown in Box 29.1 are commonplace. So as to emphasise the nature of the concept of hereditament the following are given as illustrations of what is or is part of a hereditament:

- advertising and other rights over land
- builder's huts on a site
- buildings
- fishing rights

Box 29.2 Valuations for rating of different properties

Rental method	factories offices shops warehouses workshops	
Profits method	caravan sites garages hotels leisure centres piers (for leisure) public houses race courses restaurants service stations	Typically, businesses exhibit different ways of working. Similarly, the accounts, although subject to standards, are used differently and portrayed differently. As a result, the profits method rarely has a common approach between types or classes of property.
Royalty basis	minerals (in mining a hereditament)	
Contractor's basis	airports colleges crematoria fire stations leisure centres libraries plant and machinery schools swimming pools	The contractor's basis is used frequently for buildings which are publicly owned. As a result enactments prescribe, for instance, the rate of interest which may be used.
Formula	Statutory undertakings: • canals • electricity supply • gas • pipelines • railways • telecommunications • water supply	The formula method is laid down but is, it seems, being phased out.

- land
- mines
- plant and machinery
- quarries
- structures.

In some instances a hereditament is exempt or enjoys a relief from rates (see below).

Plant and machinery

Issues arising from the long term practice of rating plant and machinery resulted in an enquiry in the early1990s and the subsequent *Wood Report* in 1993. As a result a new enactment confirmed the general principle that plant and machinery is rateable but only that specified (in the Valuation for Rating (Plant and Machinery) Regulations 1994 (SI 1994 No. 2680).

Transitional period

After the rating list comes into force there is a transitional period of five years for the adjustment of extreme changes to the level of liability a business might suffer. It works both ways:

- an occupier who would have had a large drop in the annual rates is required to pay increasingly smaller amounts each year until the new level is reached
- an occupier who would have had a large increase, pays increasingly larger amounts as the years pass.

The question of the nature of the transitional arrangements has raised concerns and the matter is being reviewed.

Exemptions and reliefs

Box 29.3 is a list of properties and organisations which are exempt or enjoy reliefs from non-domestic rates.

Agricultural land and buildings

Agricultural land and buildings were de-rated in 1929 and have remained exempt since then. However, some properties connected with a working farm may be rated — generally as a result of diversification. Typical properties caught in this way include:

- farm shops
- holiday accommodation such as caravans and cottages
- rural business accommodation (not connected with the farm), such as small factories and workshops
- sporting rights, not exercised as part of the occupation of the farm land
- stables
- studs.

Domestic property

Dwellings are exempt from business rates, but are subject to local taxation under the statutory provisions for council tax. However, composite property needs consideration when being valued (see Chapter 30). (In Northern Ireland dwellings are subject to rates, not council tax.)

Box 29.3 Exemptions and reliefs from non-domestic rates

Exemptions to property

Agricultural land and buildings	• farmers pay council tax for the farm house • other dwellings used permanently, eg cottages and caravans, are the subject of council tax • diversification businesses may result in business rates • any fish farming by a farmer is exempt from business rates.
Air-raid works	• exempt
Beekeeping	• exempt
Fish farms	• exempt
Grass drying plant	• exempt
Lighthouses, buoys and beacons	• the property belongs to Trinity House • the property must be operational
Parks	• applies to parks which are provided for public enjoyment
Places of religious worship	• includes churches, chapels and their halls
Property used for disabled persons	• adaptations not included
Sewers	• exempt

Reliefs to organisations

Charities	• properties occupied wholly or mainly for charitable purposes
Other non-profit making organisations	• can include non-profit making sports clubs and community groups
Hardship	• where without relief, a ratepayer would sustain hardship, and it is reasonable in the interest of council tax payers, the council can grant relief.
Small business relief	• from 1 April 2005, relief is available to businesses that occupy one property and the property has a rateable value of less than £10,000

Reliefs for property

Empty property	• for industrial properties, 100% relief • for non industrial properties, 100% for the first three months of vacancy and 50% thereafter
Part empty property	• provided the Valuation Agency Office agree to temporary apportionment of the rateable value • relief may be available from the billing authority

Exemption due to location

Enterprise Zones	• exemption ends when the 10-year period for a zone ends • 100% applies to commercial, industrial and hotels
Rural food shops	• relief applies in England and Wales (see the Rating (Former Agricultural Premises and Rural Shops) Act 2001)

Proposals to change the list

Once the rating list has been published it may be changed by the officers of the Valuation Office Agency in accordance with statutory procedures or Practice Statements issued by the Valuation Office Agency. Changes arise when an occupier or other person finds the assessment unacceptable or because there has been a change of circumstances affecting the property (see Box 29.4).

Future

Revaluation

The new rating lists are due to become operable on 1 April 2005. There is then a six month period in which proposals may be made to alter an assessment.

Balance of funding review

Central government is currently reviewing the way in which local authorities are funded. This may restore full or part control of the national non-domestic rating multiplier to local authority control.

Box 29.4 Rating issues arising when works are carried out to property

Construction of new non-domestic property	• check for exemption or relief — may depend on each occupier's status • likely need to negotiate with the Valuation Officer on the assessments • a liability arises when a building becomes occupied or if unoccupied, on expiry of a period of six months after the property became substantially complete • occupation by the builder of any site office or sales office or both may be liable to business rates
Self-build by the site owner	• where any person is living on the site in a caravan, it is likely to result in assessment for council tax • similarly, business operations may result in assessment and liability for rates
Alterations to a building	• change in rental value are likely to result in change in the assessment • improvements (resulting in an increase in value) result in a higher assessment on a national revaluations • negative improvements may result in a reduced assessment — make a proposal to the Valuation Office Agency
Demolition of a building	• when the building is first vacated, seek cessation of tax • partial demolition may prevent incidence of tax under vacant property charge or at least reduce the value of the property • after demolition make a proposal to remove from the list
Repairs and maintenance	• repair and maintenance should not result in an increase in value • exceptional standards of repair and maintenance, so that the property is outstanding, may increase the value over others (a possible valuation issue but unlikely)
Conversion to flats of the upper floors of a shop or offices	• the flats will become chargeable dwellings for council tax • business rating (and any composite property element) will, no doubt, change — make a proposal • planning permission will normally be required • capital allowances may be available
Conversion of a dwelling to a holiday cottage etc (or a change of use)	• takes a hitherto chargeable dwelling into business rating • business comes within the income taxation regime, eg possible claim for capital allowances

Valuation for Taxes

30

Aim

To examine the valuation of property for taxation purposes

Objectives

- **to describe the roles and activities for valuation for taxes**
- **to define value for various national and local taxes**
- **to explain the case law interpretation of the definitions**
- **to outline the appeals system on valuations for taxation**

Introduction

The capital taxes, business rates and council tax all require valuations to be carried out from time to time. Other taxes may require valuations from time to time but they are infrequent and tend to be specialised. The valuations are all statute-based and are either to capital value or to rental value (the latter for business rates).

The field of valuation for taxation is somewhat broken into sub-fields of specialised work. For instance, business rating requires knowledge of both business property and rating and the latter may require knowledge of a particular property sector. Thus, an individual may spend a working life as a rating surveyor, but as a specialist dealing with hotels and other hospitality sector properties.

Roles and activities

In Chapter 26, Box 26.1 identifies the principal roles and activities in valuations for taxation. Many professionals devote their working lives to one or more of the taxes. A valuer who specialises in appraisals for taxation must have a comprehensive knowledge of the taxation, at least in so far as taxes which relate to property.

Measurement for taxation

The RICS Code of Measuring Practice 5th ed shows under applications the measurements which might be used for business rates and council tax. Box 30.1 summarises the approach to measurement in a very broad manner.

Box 30.1 Measurement for business rates and council tax — summary, based on the RICS Code of Measuring Practice

Measure	Tax	Type of property
Effective floor area	council tax	flats and maisonettes
Gross external area	council tax business rates	houses and bungalows industrial (Scotland) warehousing (Scotland)
Gross internal area	business rates	food superstores industrial (England and Wales) special property (based on cost) warehousing (England and Wales)
Net internal area	business rates	business use (other than elsewhere above) composite property offices shops supermarkets

---

Bases of valuation

Valuations for taxation are statute based. The precise wording of the definitions of capital value and rental value for the main taxes affecting property is explored below. The assumptions for each definition are explained below. A statutory basis does not necessarily mean that the valuer has a clear remit for determining value in a given case. Case law and mandatory or discretionary professional valuation practices and standards need to be taken into account.

Statutes

Being statute based has meant that each definition is likely to have been the subject of interpretation and negotiation. This has often resulted in disputes which have been settled by the Lands Tribunal or the higher courts. A measure of the case law on valuations is therefore given in Box 30.3 below.

Professional practices and standards

Although the valuation and appraisal standards mentioned below apply to valuations for company reports and financial statements, generally, they include guidance which may be helpful for valuations for taxation.

In the UK the Royal Institution of Chartered Surveyors (RICS) and the Institute of Rating, Revenues and Valuation (IRRV) are the main bodies concerned with the practice of professional valuations. The principal material from professional bodies associated with valuations is:

- RICS Appraisal and Valuation Standards 5th ed (May 2003)
- RICS Code of Measuring Practice (November 2001) (see below).

Capital taxation: open market value
Definitions

The definitions of open market value for IHT and CGT are the same but different assumptions apply to each, eg the date of valuation. Much of the case law interpretation arose from early estate duty cases but, seemingly, their validity holds. Nevertheless, a valuer should consider the circumstances of the valuation to hand against those of an earlier case — there may be differences. So as to give insights to the concept of open market value Box 30.3 gives a few of the nearly 20 cases which are readily identifiable on the topic.

Date of valuation

As in all valuations, the date of valuation, ie the date of the event for which the property is to be valued, is specified in the relevant statute. The various dates are considered in Box 30.2 for each tax.

Inheritance tax (IHT)
Definition and case law

S160 of the IHT Act 1984 defines open market value for IHT, namely:

> ...the price which the property might reasonably be expected to fetch if sold in the market at the time of the death...

Many cases may be cited to give perspectives of the difficulties that have arisen in the past in interpreting the wording of the definition (which goes back to the definition in the Finance Act 1894) (see Box 30.3).

Statutory assumptions are required in particular situations — mainly where reliefs are available. Box 30.4 gives the flavour of the context and the nature of the assumption the valuer must make. The reliefs are dealt with below, ie the box only refers to the valuation assumption.

Box 30.2 Dates of valuation for taxation

CGT	• for a sale, normally, the date of contract • for a conditional contract, when the condition is satisfied • for very old acquisitions, 6 April 1965, perhaps with time-apportionment (realistically, probably not operable now) • for re-basing, 31 March 1982 (replacing time-apportionment on very old acquisitions) • allowable cost of improvements before 31March 1982, within the market value at that date (they are rebased and cost becomes value) • for a gift, the date of gifting
IHT	• the date of death • gifts in the seven years before death (potentially exempt transfers), the date of death • for a sale from and within two years of the death at a price lower than the probate price, the date of the sale (the price is substituted for the probate value)
Council tax	• for the current valuation list, normally 1 April 1998 (but see text) • for the current revaluation, 1 April 2003 (but see text) • for alterations, the date of a subsequent sale (bearing in mind 'tone of the list') • for demolished buildings, the date of proposal • for new buildings, the date of completion
Non-domestic rating (business rates)	• for present rating list, 1 April 1998 (but see text) • for the current revaluation (and appeals for it), 1 April 2003 (but see text) • for demolished buildings, the date of the proposal • for new buildings, the date of completion
SDLT	• the new regime is designed to prevent avoidance or delays in paying the SDLT; there are, therefore, many possible dates depending on the circumstances of each or a series of transactions
VAT	• at the point of sale for VAT (date of contract for land, etc)

Capital Gains Tax (CGT)

Definition

CGT arises when a person makes a disposal. In many instances there is no need for a valuation since the property is sold at open market value. Gifts, exchanges, sales at less than open market value (under value) and certain other disposals, ie where there is no consideration or the consideration is at less than open market value, mean that an estimate of the open market value will be required.

Valuations

Valuations for CGT are assessed to capital value on an open market value basis in accordance with a definition which is very similar to that given for inheritance tax.

Box 30.3 Cases on the interpretation of open market value for capital taxation

Prudent lotting	*Earl of Ellesmere* v *IRC* (1918) and *Duke of Buccleugh* v *IRC* (1967) An estate should be (hypothetically) lotted in such a way as to achieve the highest possible price
Special purchaser	*IRC* v *Clay* (1914) The open market value could be the price that a special purchaser might be expected to pay, eg a sitting tenant.
Sales after death	*Middleton and Bainbridge* v *IRC* (1954) Sales to sitting tenants shortly after death were the best evidence of the value at the time of death
Costs and fees of sale	*Duke of Buccleugh* v *IRC* (1967) No deduction should be made for the hypothetical costs and fees of the assumed sale by the vendor.
Vacant possession value — investment property?	*Harris* v *IRC* (1961) Vacant possession value not appropriate for the furnished letting — make an allowance for obtaining vacant possession at maximum difficulty. *IRC* v *Graham's Trustees* (1969) Freeholder owned a farm which was let to a partnership of which he was partner. Partnership ceased on death. Held that the vacant possession value was the open market value.
Arrangements for sale	*Duke of Buccleugh* v *IRC* (1967) Assume that the arrangements for the hypothetical sale are made before the date of death.
Best price against highest price	*Re Haye's Will Trusts* (1969) Deceased's farm was agreed at a value of £48,000 and sold to deceased's son (in accord with the will). Sisters disputed the value as being too low. Held that the best possible price did not mean highest possible price — there is something like a plus and minus 10% around an agreed figure.
Time or date of valuation	*Duke of Buccleugh* v *IRC* (1967) The date of death is the date of valuation.
Valuer's estimate, actual price	*Crossley* v *IRC* (1954) Valued at £4,500 at the date of death by a valuer but later offers of £5,000, £6,000 and finally £7,000 made six months later. Valuer's estimate accepted.
Restrictions on the sale	*Duke of Buccleugh* v *IRC* (1967) Actual restrictions preventing an open market sale should be ignored.

In general, an interpretation of the definition for CGT follows the cases given in Box 30.3 above for IHT. The period of accrual for a gain (or loss) for CGT passes has changed from time to time. However, most cases relate to the period from or a discrete period starting since 31 March 1982 (when CGT was rebased from that date).

Box 30.4 Valuations for IHT requiring assumptions

Large estate	• generally, any notional 'flooding of the market' by putting it on the market at the time of death, any resulting decrease in value should be ignored
Agricultural property	
• without development potential or other high value features	• value as agricultural property
• with development potential etc	• first ascertain the open market value
	• second, ascertain the value with a perpetual covenant that the property remains agricultural (see s 115)
Stud farms for breeding and rearing horses	• treated as being agricultural property
Short rotation coppice	• treated as being agricultural property (see s 154 of the Finance Act 1995)
Woodlands and timber	
• where woodlands relief operates	• on death, the net value of the timber will be required and left out of account
• where agricultural relief operates but not that for woodlands	• woodlands ancillary to the agricultural usage, treated as agricultural property
Gift of a property	
• eg part of the garden of the family house	• the reduction in the value of the estate is required

---

Income taxation

There is little call to use a professional's knowledge and skills in income tax. At one time the praire value of land used for woodlands was required but this has not been needed since the legislation providing for the assessment of woodlands to income tax under schedule B was repealed.

Capital allowances

Aspects of capital allowances call for values to be apportioned but these are usually related to costs of construction.

Value Added Tax (VAT)

For property transactions subject to VAT, money is the usual consideration passing and will, therefore, be the amount charged to tax. However, in a small proportion of transactions, a property may be sold in exchange for one or more properties, cash in part or other non-money assets. Where the exchange property involves land and buildings, a valuer will need to appraise the value.

Business rates

In valuing business property for non-domestic rating there are six main pointers which the valuer has to address, namely:

- the definition of value which needs to be found
- the assumptions underpinning the valuation
- the state and qualities of the property to be valued
- the state of the locality in which the property is situated
- the date with regard to which the valuer must consider the locality
- the date of valuation (which will not necessarily be the same as the previous date).

In a somewhat naive summary, the valuer wants to value the property at a given date, in a locality as it may be at another date or (the same date) in a state in which it may not actually be the same as it is or was at the first or later date ...!

Since 1 April 1993 business or non-domestic property has been revalued or assessed at five yearly intervals to rateable value, an annual net rental value. Although the date of valuation has been or is 1 April in year X, (... 1998, 2003 ... etc) the locality is assumed to be in the state that it is expected to be on 1 April in year X+2 (... 2000, 2005 ... etc). The two common dates for valuation and for the locality respectively create the basis for fairness between the ratepayers, commonly called tone of the list. Subsequent proposals to alter an assessment must have regard to the concept. If the property has been altered in some way but the locality has not altered there is no problem but if the property has altered and the locality has altered by the date of proposal, the valuer faces a different situation to that of the original date. The legislation, seemingly, provides the basis to deal with it! In practice the date for values is the antecedent date, eg 1 April 1998 for the list of 1 April 2000. For the property, it is the state of the property at 1 April 2000 (in this example) or the subsequent date of the change to the property.

In principle, the rating of such property in this way goes back at least 400 years to the Poor Relief Act 1600. Case law has refined the professional valuer's understanding of the definition and may still be important in usage even though legislation has attempted a codification of it. In fact a reading of the cases, or summaries thereof, is a good way to begin to understand the nuances of how a valuer does or should deal with valuation work in general.

Council tax

Valuations for council tax are banded so there may not be the pressure to appeal against an assessment that might otherwise occur, ie without banding. Although the valuation of dwellings is the responsibility of the Valuation Office Agency, in the last revaluation (being the first for the council tax) the work was mainly undertaken by valuers in the private sector. They worked, of course, to the statutory definitions and assumptions — the date of valuation was 1 April 1991 (locality at 1 April 1993).

Valuation assumptions

The capital value (open market value) of a dwelling for council tax has the following assumptions:

- the property is on the (notional) open market with vacant possession

- the property is freehold, except in the case of a flat when it is leasehold for a term of 99 years at a nominal rent, say £1 (but this is not specified)
- the property is being sold free of any rent charge or other encumbrance
- the size, layout and character of the dwelling and physical state of the locality were the same as at the relevant date
- the property is in a reasonable state of repair
- where the occupier enjoys the use of common parts, they are in a reasonable state of repair and the purchaser contributes to their upkeep
- fixtures for disabled persons which add to value are not included
- use would be permanently restricted to use as a dwelling
- there is no development value, other than value attributable to permitted value.

The source of the assumptions is the Council Tax (Situation and Valuation of Dwellings) Regulations 1992 (SI 1992 No 550). The problem of tone of the list is dealt with in a similar way to that for business rates.

Composite property

A composite property is one which is part business and part domestic. In valuing composite property two approaches are conceivable, namely:

- use direct comparison — analyse sales in the locality and apply the outcome to the subject part property
- value the whole property — then apportion the value to the two parts, ie the non-domestic and the domestic.

In *Salvation Army* v *Lane* (1994) the listing officer used direct comparison and the Salvation Army adopted the other method. The Lands Tribunal adopted the latter approach with the result that the assessment was much lower.

Appeals

In the first instance a council tax appeal against an assessment on the valuation list is made to the Valuation Tribunal and hence to the High Court. The Lands Tribunal hears appeals from the Valuation Tribunal on rating matters. Its jurisdiction therefore includes the following kinds of valuation:

- capital gains tax
- business rates
- council tax
- inheritance tax.

Appeals from the Lands Tribunal go with permission to the Court of Appeal and hence the House of Lords. Chapter 34 examines the role of the Lands Tribunal in resolving valuation disputes between the Valuation Office Agency and the payers of rates, council tax or national taxes.

Part 9

Property and Government

Government and Property

31

Aim

To explain the "government" involvement in the property industry

Objectives

- to give an account of the structure of government
- to briefly show how politicians are elected
- to describe the formation of a policy and its passage to enactment
- to outline government policies affecting the property industry

Introduction

The ways in which government affects property is explored by examining the range of powers and broad functions of international, national, provincial and local government in the UK. Although parliament is the starting point, being the font of almost all enactments affecting property, other bodies also participate in law making. This chapter, therefore, looks at the structure of government at all levels.

The next chapter examines the roles and functions of government departments, departmental agencies, non-government public bodies and other public bodies, as well as local authorities.

Structures of government

Apart from the national government there are several other levels which may result in property-oriented policies. The levels are the EU, the Scottish Parliament, the Welsh Assembly and local government at regional, county or unitary levels, district and town or parish levels. Each has functions which deliver policies, programmes and projects, as shown in Box 31.1.

International

The EU treaties oblige the development of a pan-European economy which provides:

Box 31.1 Distribution of selected governmental functions in the Great Britain

Parish and town councils (rural England) community councils (Wales)	• councillors • parish clerk	• local amenities • footpaths • playgrounds
Metropolitan boroughs district councils (country areas)	• councillors • cabinet members • director	• planning functions • housing strategy • waste management
County councils	• councillors • leader and cabinet members • director	• mineral and waste planning • education • highways
Unitary authorities	(as county council)	• all functions of local government
Regional bodies (England) London Assembly	• representatives • elected members • elected mayor and elected members	• economic and social planning • regional housing strategy • transport
Scottish Parliament Welsh Assembly	• elected members	• devolved policies • planning law
Parliament	• prime minister • ministers	• all function of national government (see Chapters 32 and 33)
• House of Commons	• leader • members of parliament	• scrutinise legislation • vote on Bills
• House of Lords	• members	• monitor government performance • represent constituents
European Parliament	• members of the European Parliament	• scrutinise policy • monitor legislation • approve European Commission members
Council of Ministers	• President • Ministers of member states	• formulate policy guidelines for European Commission to progress • oversee the work of the European Commission directorates

---

- freedom of movement of capital
- freedom of movement of persons
- freedom of movement of goods
- freedom of movement of services.

The UK's membership of the EU has resulted in a substantial body of EU law, being incorporated into UK law, much of which directly impinges on property, eg VAT.

Much of the EU's work is done through the directorates of the European Commission. Each of the 20 or so directorates, headed by a directorate-general, is responsible for a function of government, eg energy, environment.

Devolved government

Partial self-government for both Wales and Scotland has been introduced in recent years. They have different forms of self government and there are some implications for the property industry. Both the Scottish Parliament and the Welsh Assembly have powers to develop different policies to those created for England by Parliament; with concomitant powers to raise a limited level of taxes.

Regional

The transition to region-based government is a recent phenomenon in the UK. The approaches in England, Scotland and Wales are somewhat different. Some governmental powers have been shaved off central government and others are derived from local government. The traditional functions of local government have changed with the introduction of the regional structure and further developments may arise. The general principle seems to be to operate functions of government at the level nearest the citizen — subsidiarity.

Quasi-government

In the past a government has created a quasi-government body to address such concerns as the distribution of population or the regeneration of an area which has declined. The subject-matter may be one which no individual local authority could tackle or one which extends beyond the boundaries of one local authority. Examples of the concerns, in terms of the solutions are as follows:

- the new towns development corporations — set up to create 30 or so new towns to shift populations from congested conurbations
- the London Docklands Development Corporation — created to address the abandonment of the East London docks and the need for regeneration
- national parks to plan and manage areas of beautiful countryside
- the urban development corporations set up under the Local Government, Planning and Land Act 1980.

Under the New Towns Act 1954 and later Acts, the development corporations had powers akin to local authorities, such as:

- making compulsory purchase orders to acquire land
- clearing land and installing infrastructure
- developing serviced land
- managing property
- selling and leasing land and buildings.

Later, the same kind of model was used to deal with the East London docklands, namely the London Docklands Development Corporation and the urban development corporations, Thames Gateway.

National parks are other bodies which make a governmental impact on settlements and individual properties or larger estates. Their concern is to preserve and maintain the idyllic environments within their charge. The essentially rural perspective they are beholden to tends to be restrictive of new developments which are likely to mar the countryside, eg quarrying or other industrial businesses which want to locate or start-up there.

Policy into legislation and practice

A broad overview of the numerous policies, programmes and projects introduced and developed (or rejected) by successive governments are given in Box 31.2.

Parliament

Traditionally, in the UK, following national elections, the government is formed from the political party with a majority in the House of Commons. Thus, at about five-yearly intervals voters elect a member of parliament (MP) for their constituency. Most of the 600 or so MPs elected to the House of Commons are put forward by one of the major political parties with a remit to promote their party's election manifesto. The manifesto contains the promise of policies on most matters, including property.

In effect the party with a majority of MPs forms a government under the Prime Minister, who in turn appoints ministers, mainly from senior MPs or from members of the House of Lords, to run the departments of government, eg the Department of Trade and Industry. The ministers will take the manifesto's promises and convert them into a five-year programme of proposed legislation or government practices (see Chapter 32).

EU legislation

In addition to UK policy formation, ministers take part in the formation of policies at the EU's Council of Ministers (in effect several "councils" — for each function at the European level). The European Commission develop the policy for approval by ministers after it passes through the EU's structure. Any directives resulting from this process must be incorporated into UK law — by passing though the UK legislative processes. Legislation emanating from the EU in recent years included the following statutes which directly affect the interests of the property industry in the UK:

- the value added tax regime
- the EU procurement rules for big projects
- the regime for landfill taxation
- the system for waste management
- the regime for the protection of the workforce
- the carbon emissions trading legislation
- the land contamination regime.

Box 31.2 Major controls, programmes and projects introduced by successive governments

1940s
- planning system of controls
- attempt to tax development value, with development charge
- clearance areas initiated to deal with bomb damaged towns

1950s
- slum clearance in major towns
- new towns programme initiated with new town development corporations

1960s
- urban regeneration replaced slum clearance
- compulsory purchase was 'rejuvenated' and codified
- motorways programme commenced
- railways coverage reduced
- corporation tax introduced
- capital gains tax introduced
- attempt to tax development value, with betterment levy

1970s
- water industry opened to leisure access by the public
- forestry opened to leisure access by the public
- capital transfer tax replaced estate duty
- attempts to tax development value with development gains tax and development land tax
- Community Land Scheme attempted
- UK joined Europe
- VAT introduced

1980s
- privatisation commenced
- freeports were introduced
- inheritance tax replaced capital transfer tax
- enterprise zones introduced with a 10-year life
- consolidation of income tax and corporation tax

1990s
- privatisation continued
- urban development corporations started
- consolidation of capital gains tax
- council tax introduced
- business property to national non-domestic rate system
- build up of environmental taxation, landfill tax, aggregates tax
- joined-up government initiated
- government for Scotland and for Wales
- regional government initiated

2000s
- planning system and compulsory purchase system overhauled
- capital allowances consolidated
- consolidation of value added tax
- land registration overhauled
- constitutional affairs overhauled
- building regulations extended to sustainability and security

Something like 20,000 pages of legislation has been promulgated by the EU. The purpose, in general, is to create a level base for the common competitive market. Much of the legislation has been, seemingly, cherry-picked from the mass of laws at member state level. For instance, at least two countries have or will have a levy on consumer goods (to pay for the EU recycling legislation which is being adopted by member states). The levy will probably become a common policy in due course.

Management, operations and change

The day-to-day management of the government's policies is effected by civil servants. New policies which require legislation are promoted primarily as draft legislation — Bills, but this is not the only approach to change. For instance, manifesto promises and other policy proposals for change have been, but not invariably, announced by a consultation Green Paper which is distributed to interested parties for advice or comment. After a consideration of the responses by the government (ministers and their advisers) an adopted policy will be published for comment as a White Paper (indicating a firm policy) or perhaps as a parliamentary Bill. Further scrutiny of the draft policy, ie as a Bill, (and tweaking of the legal wording of its content) takes place in committee stages by members of both Houses. The pre-legislative stage is very important in that it allows pressure groups and others to lobby, make representations and discuss the draft with ministers, members and civil servants. It is the fine tuning stage which is critically necessary — and mitigates the need for litigation.

Another approach adopted in the UK is the from the research work undertaken by the Law Commission. Bodies of law which are causing concern are selected, researched and reported on by the Law Commission. The work usually leads to a consultation process; and often ends with a new Act being processed by parliament for royal assent.

Parliamentary procedure

All levels of government are required to act within their powers as set by national legislation. As mentioned above, both Scotland and Wales have devolved powers in some areas of government. Also, several Bills are promoted each year by local government authorities for projects and programmes of a local nature. Over the years, having become Acts, many will have had and continue to have implications for the local property market.

The main areas of policy developed by successive governments have been explored in earlier chapters. In the main the presentation has been on the current policy of the present government, covering new policy formation, eg aspects of building control, compulsory purchase, land registration and planning, in the context of the body of the law on each topic. On occasions historical aspects of a policies development have been highlighted. The sections which follow draw on the earlier material and link it to state-of-play perspectives on policies affecting the property industry. Some detail is given in Chapter 8 on the sectors of the property market.

Policies affecting the property industry

Agricultural policy

International trade in agricultural products and developments in agriculture in the EU have made it increasingly difficult for the occupier of the smaller unit to obtain a living by traditional modern farming unless diversification of land use is permitted (see Chapter 8). Diversification has, therefore, been developed as a rural policy so that many non-rural type developments, commercial or residential, are now permitted on a scale which fits the usual countryside scene, eg redundant agricultural buildings are allowed to become industrial workshop units. Many barns and other agricultural buildings, eg oasthouses of Kent, which have been converted to residential use, reflect the policy towards diversity and support for farmers. Energy policy is bringing new crops to agriculture, eg short rotation coppicing for biomass electricity generation. This trend is also reflected in the new kind of tenure which was introduced by the Agricultural Tenancies Act 1995, the farm business tenancy.

Construction

The construction industry has felt the impact of new policies, including:

- emphasis on the use of recycled materials
- health and safety legislation
- pressures for the adoption of sustainable construction methodology, eg the Building Research Establishment's eco-Home initiative
- the return of apprenticeships.

Proposals to change the education system may result in vocational training in construction being offered in schools.

E-policy

The government is committed to developing electronic means of communication for many aspects of the civil service work. The numerous examples of electronic technology have been or are being adopted by government departments and agencies may be cited. Those matters which impinge on the property industry include:

- the many statutes available electronically on the Internet
- the Land Registry is empowered under the Land Registration Act 2002 to develop e-conveyancing
- electronic property information for house purchasers and others by various agencies, eg the Environment Agency
- proposals to change the rating assessment of business property may be made electronically to the Valuation Office Agency
- property information from a variety of government sources is available in on the Internet, eg all departments have home pages
- individuals may complete their tax returns to their inspector of taxes;

Local authorities are also facing information and communication technology challenges, with it seems encouragement from government. Examples are:

- electronic mapping by satellite — for accurate property boundary identification
- wireless local area networking (WLAN) — for urban street activity monitoring and control, for staff to access and send in data while outside, and for resident input with local data

- electronic systems — for collecting council tax and remedying common problems in a responsive manner.

Energy policy

Renewable energies are to the forefront of government policy on energy. The targets are some 10% and 20% of energy from renewable sources by 2010 and 2020 respectively. Special incentives have been introduced to encourage generators to create electrical power from renewable sources and suppliers are required to buy a proportion of energy from those sources.

Energy management is also encouraged in homes and other property sectors by means which include:

- grants for solar energy systems in homes
- the climate change levy
- trading in carbon dioxide emissions.

The building regulations have been amended to increase the emphasis on energy conservation in new and existing buildings.

Environmental policy

A great deal has been done to improve the environment, in most property sectors. Much of the thrust has come from the EU. As a whole the policies include:

- environmental impact assessment for major developments
- the regime for the remediation of contaminated land
- the regime to improve the nation's performance on waste and waste management
- the development of green transport
- improvements to the water supply, eg the campaign against leakages in the supply side
- standards for beaches and sea water
- the introduction of energy taxes, eg aggregates levy, landfill tax and the climate change levy
- the emphasis on sustainable settlements and developments.

Forestry and woodlands policy

Briefly, forestry policy has changed since the Forestry Commission was charged with developing a strategic reserve of forests for timber production. The emphasis has changed to:

- disposals of forests and woodlands to private commercial growers or to conservation bodies like the Woodlands Trust
- increased public access for amenity and recreational purposes has developed since the 1970s
- planting is now more considerate of the local typography and natural ambience of the countryside, eg a policy of deciduous planting is replacing conifer plantations
- tax changes so that inappropriate planting is eliminated or at least reduced.

Housing policy

Nationally, the Office of the Deputy Prime Minister has indicated at least 450,000 new homes will be required to cope with increased household formation over the next few years. Each local authority has been required to develop a housing strategy for its area indicating the probable need for housing in the foreseeable future and the housing services to be provided. However, it seems that there is a gap in the new planning regime in that the relationship between planning strategies and the housing strategies is not clear. Also, there is some indication that the regional authorities might be empowered to deal with housing at the regional level.

The private sector has been relatively healthy in recent years. Concerns on policy have arisen, including:

- the emphasis on planning and development of brownfield sites and concomitant lack of planned releases of greenfield sites
- the use of s 106 agreements to secure affordable housing on private building sites
- the lack of keyworker or affordable accommodation in places where the level of values is high.

Social housing is a relatively small part of the market. Its origins are in the 19th century housing provision by voluntary bodies like the Peabody Trust. This was followed by over half a century of housing being provided by local housing authorities, roughly until the late 1960s and early 1970s. Since then new building by local housing authorities declined substantially and the social registered landlords replaced them; the social registered landlords are mainly housing associations funded by the Housing Corporation and building societies.

Provision by social registered landlords replaced that of many local housing authority housing estates, eg West Kent Housing Association acquired Sevenoaks District Council's estates. Although social registered landlords still build new houses and acquire existing buildings for conversion, the provision has become less intense. Present emphasis by government is on the provision of affordable housing for the relatively low paid in towns and cities and in rural areas where local young people find it difficult to obtain a dwelling, sometimes in partnership with private developers as a result of s 106 agreements.

Industrial and commercial policy

The decline of heavy industry, dockland industry and manufacturing industry, together with the forecast slowing down of the oil and gas industry has required government to address the planning issues. Policies have included:

- special grants and support for particular industries in decline, eg the coal industry
- encouragement of investment with a changing range of capital allowances
- the development of special area status, eg business improvement areas, enterprise zones, freeports and simplified planning zones
- adoption of the EU assisted areas policies for Cornwall, Northern Ireland and other relatively distressed areas.

Retail policy

The recent past has seen the advent and rapid growth of provision of out-of-town superstores, outlet centres and retail parks. As a result the government has become concerned about the impact on town centres. Although the national retail market has been said to be becoming saturated, provision seems unabated. Two possible concerns for the property industry are changing patterns of demand due to:

- the changing formats — petrol filling stations introducing local shops
- internet shopping — consequential distribution patterns should this take-off.

It seems that town centre management may need to address the issues of decline due to out-of-town shopping centres, eg regional shopping centres, like Bluewater or Lakeside on the Kent and Essex sides of the Thames Estuary respectively, draw thousands of shoppers daily.

Taxation policy

From time to time consolidations of bodies of taxation statutes take place. In the meantime changes are introduced for the following reasons:

- to meet perceived abuses of the system by evaders
- to stem revenue losses due to the adoption of devices for tax avoidance, particularly from the seemingly artificial or creative avoiders
- the adoption of a new tax with a particular policy thrust, eg the many environmental taxes.

Transport policy

Property development reflects a trend towards a greening of many aspects of transport policy, including:

- a planning application for major new development requires the joint submission of a green transport plan
- grants and tax relief to motorists for cars using liquid gas — leading to the need for liquid gas supply points
- tax incentives to those who use cycles for travel to work — slowly leading to workplace or publicly provided facilities for cyclists, eg changing rooms and cycle racks
- the workplace parking levy is intended to reduce travel to work by car
- requirements for car parking have been reduced for new development, with the view to encouraging the use of public transport
- developments for park and ride facilities — to ease traffic congestion
- congestion charging affecting retail businesses — introduced to ease congestion in towns and cities and to raise funds for improvements to public transport.

Planning policy

Many planning policies or policies which have planning implications are referred to in the sections above. The changes provided by the recent overhaul of planning in the Planning and Compulsory Purchase Act 2004 are being introduced; so the system is in transition. The emphasis is to embody sustainability in new development and in the culture of planning.

Performance and best value

Within the government structure, the measurement of the performance of public bodies is effected by means a review and assessment of best value performance and plans. Essentially, standard performance indicators are published by the Audit Commission as a basis for setting targets and assessing the performance of a public body against its targets; and against a best value authority's peers. The performance measures cover much of the public sector and include such responsibilities as:

- care of the environment
- fire services
- housing provision
- land searches
- planning searches
- police
- waste disposal.

It seems that in Wales nearly 90 performance indicators are used as measures of local authorities. However, the Welsh Assembly is intending to reduce the number to less than 25%, ie 21 performance indicators.

An illustration of the procedure adopted by the Audit Commission is given in Box 23.3 in Chapter 23.

Pressure groups

Although mainly outside of government, many organisations advise or pressure governmental bodies to adopt, amend or drop policies, programmes or projects. On any concern involving property, many managers, consultants and other professional advisers are likely to be working on one, two or more fronts. A few of the possible pressure groups for an individual dealing with policy implications for the property industry are shown in Box 31.2.

There are of course hundreds of representational groups who may make representations on professional, trade or other matters, including:

- advising the senior management of possible implications for the business of the proposals
- advising clients of the opportunities and threats arising from the proposed changes
- seeking changes to the proposals by representing issues, problems, hardships and injustices to trade associations, professional bodies and to government officials
- drafting amendments to any Bill to allay inconsistencies, errors and omissions, eg by promoting exemptions and reliefs
- if politically active, making representations directly to ministers or opposition speakers or by other means.

Box 31.2 Pressure groups related to property

Architecture	• Commission on Architecture and the Built Environment • Georgian Society • Victorian Society
Business	• Institute of Directors • Confederation of British Industry • Agriculture and Business Association
Environment	• Council for the Protection of Rural England
Local government	• Local Government Association
Planning	• Royal Institute of British Architecture (RIBA) • Royal Institution of Chartered Surveyors (RICS) • Royal Town Planning Institute • Law Society • Town Planning Association
Construction	• Chartered Institute of Building (CIOB) • RIBA • RICS
Agriculture	• Agriculture and Business Association • RICS
Built Heritage	• Historic Houses Association • National Trust
Forestry	• Forestry Contractors Association • RICS
Housing	• Chartered Institute of Housing • Building Societies Association • House Builders Federation

Government and Departments etc

32

Aim

To examine the relationships between the property industry and government and government departments

Objectives

- **to describe the main functions of government**
- **to identify property-related departments, agencies and other government bodies**
- **to show the main property-related programmes and projects of the various government bodies**

Introduction

The property industry faces many government and quasi-government bodies which have responsibilities to regulate, control or perhaps assist those in the industry. This chapter will survey the how property interests are affected by government. Where appropriate it will indicate where a particular topic is dealt with in greater detail elsewhere in this volume.

Roles and activities

Box 32.1 gives the roles and activities of the ministers involved, in general, with property industry matters. It is intended as an insight to the positions of ministerial appointments, rather than property industry professionals and others employed in government. The juxtaposition of topics indicates an appointment of a minister of state (*) or a parliamentary under-secretary of state or an equivalent position(^).

Box 32.1 Main ministerial appointments and duties *vis-à-vis* the property industry

Deputy Prime Minister (ODPM)	• local government and the regions (*)
	• regeneration and regional development (*)
	• housing and planning (*)
	• sustainable communities, social exclusion and homelessness (^)
	• regulatory reform and fire safety (^)
Chancellor of the Exchequer (HM Treasury)	• finance (^)
	• economics (^)
	• treasury and private finance initiative (Chief Secretary to the Treasury)
	• tax issues (^)
Secretary of State for Trade and Industry	• trade (*)
	• e-commerce trade (*)
	• women (*)
Secretary of State for the Environment, Food and Rural Affairs	• environment (*)
	• rural affairs (*)
	• fisheries and animal welfare (^)
	• farming, food and sustainable energy (^)
Secretary of State for Constitutional Affairs	• criminal justice and elections (^)
	• legal aid and civil justice (^)
	• family justice and tribunals (^)
	• Scotland (^)
	• Wales (^)
Secretary of State for Culture, Media and Sport	• sport (*)
	• arts (*)
	• tourism, film and broadcasting (^)
Secretary of State for the Home Office	• crime reduction, policing and community safety (*)
	• criminal justice system (*)
	• citizenship and immigration (*)
	• drugs and Europe (^)
	• race and community cohesion (^)
	• prisons and probation (^)

Functions of government

Prime functions

The government has a relatively discreet set of functions which directly or indirectly affect the property industry. Briefly, they include the following:

- law-making — creating enactments as a result of the adoption of policies which affect property directly or indirectly
- planning — creating a national planning framework to cope with the need for accommodating change with infrastructure and buildings

- funding — allocating funds to programmes and projects that are required to support policies
- taxing — raising funds by devising and operating a tax base which meets needs and accords with broadly acceptable principles of taxation.

Joined-up government

A principle which is beginning to pervade government activities on projects and programmes is joined-up government. It is intended that the various departments, agencies and other national and local bodies act in concert. In a sense it involves the principles and practice of project management at two levels at least, namely:

- an organisational planning period (before the start of the programme or project time) — when the various bodies involved take cognisance of the need to marshal resources for the later commencement
- once the project starts, the careful coordination of the various contributions the bodies will be making.

The activities before the project starts are akin to the pre-contract period in construction — the contracts and partnership agreements are drawn up, the i's are dotted and the t's are crossed and the signing is completed. Of course the essence of joined up government is not legalistic but of the spirit — the spirit of open cooperation. An example of joined up government might be the way in which numerous bodies are cooperating to deal with anti-social behaviour. Many areas have been selected for the action by local authorities, courts, police and others.

Courts and tribunals distinguished from government

Although the courts and tribunals may be thought to be part of government, under the largely unwritten constitution they are a separate organ of the state — independent of ministerial or political influence. This reflects the principle of the separation of powers in the running of the UK and the independence of the judiciary.

Case work and judicial review

Whereas Parliament makes the law, the courts and tribunals interpret the law. The bulk of court work is to settle disputes between parties, eg a breach of contract or some other matter. The latter may involve a government department or a local authority, eg the Lands Tribunal is asked to settle compensation or a point of law in a case of compulsory purchase. However, where the issue is, say a minister's making of a decision or, say, a public body's conduct of a process or procedure, the court may be asked to determine the matter by way of judicial review.

Departments of state

The government departments concerned with matters affecting property are indicated in Box 32.1. Others may have an involvement with property but only as a periphery function. Of course, all departments of government accommodate staff in all manner of property which is either freehold or held on lease. In the past some have been at the forefront of developing systems of property and facilities management which have cascaded into the public domain.

A department's other main duties for the functions it deals with include:

- advising their ministers
- taking and seeking advice on policy proposals, including proposals for EU policies
- drafting enactments, including those for EU directives on government policy initiatives
- human rights
- the implementation of EU driven enactments
- implementation of the laws of freedom of information
- judicial reviews affecting the department's work.

Treasury

The Treasury is the hub of the government's fiscal concerns, namely:

- to raise funds by taxation, borrowing or other means
- to allocate funds to departments and others
- to ensure that the money is well spent
- to monitor and control the economy.

A government's first life is nearly five years from being elected on the party manifesto, which sets out its aims and policies for the country. Implicit in the manifesto are a variety of programmes and projects which, if adopted by the government, will probably require legislation, and funding on a planned basis over the five years. The Chancellor will need to manage the economy so as to meet planned expenditures; which will, originally, have been included in the planning phase on the basis of the Chancellor's expectations for the economy. Each year the departments will have to negotiate with the Treasury for funds to meet its targets. Each year the Treasury will want savings and best value performance.

The Chancellor's annual budget cycle covers, broadly, three periods:

- from summer's end to the budget review statement in late fall (say, November)
- from late fall to spring's budget statement (say March or April)
- from spring to mid-summer's royal assent of the Finance Bill (say, July).

Each year the departments make their bids for funds and agree them with the Treasury. The Chancellor announces government expenditure, ie total managed expenditure (TME) in the budget. (For 2005–2006 it is estimated at £548 billion.) The TME is split into:

- departmental expenditure limits (DELs) (say 60%)
- annually managed expenditure (AME) (say 40%).

Any DEL not used in the year may be carried over to the next year but not that in respect of AME. Thus, each department knows how much of its earlier bids will be available, together with any DEL carried over from the previous period.

Also, from the budget, outsiders, eg property owners and professional advisers, will learn of forthcoming new taxes, taxation levels and new or discarded exemptions and reliefs from taxation. Similarly, organisations will learn of new expenditure levels for their field and other matters affecting them, eg registered social landlords may expect £20bn for new build housing in the three years to 2007–2008.

Office of the Deputy Prime Minister (ODPM)

The ODPM (and DEFRA) are the twin hubs of the government's involvement with the property industry (in the broadest sense of the term). Box 32.2 sets out each area covered by the ODPM together with a brief outline of some of the programmes and projects. There are also numerous working parties and consultative groups which have been sponsored by the ODPM; almost all are embedded in the work of the property industry.

Box 32.2 Main functions of the Office of the Deputy Prime Minister

Regions	• developing regional government in England • elections for regional assemblies (RAs) in the North East • proposed that RAs produce regional housing strategies
Local government	• transitional arrangements for rating
Planning	• major changes to the planning systems in England and in Wales (Planning and Compulsory Purchase Act 2004) • policies on sustainability • revisions to the planning guidance • revisions to compulsory purchase • development of the framework for various kinds plans • planning inspectorate
Housing	• linking regional planning and housing strategies • reform of the home buying process
Fire	• fire service • fire safety
Regeneration	• urban development corporations, eg Thames Gateway • business improvement districts
Social exclusion	• funds for local initiatives, supporting deprived areas and groups, eg breakfast clubs, vocational skills courses
Neighbourhood renewal	• regeneration projects in urban areas • major countryside parks

---

Department of Constitutional Affairs

The Department of Constitutional Affairs was set up in 2003 when it was proposed to abolish the position and office of the Lord Chancellor but this may not now happen. Nevertheless, the powers of the Lord Chancellor are being changed.

Box 32.3 Main functions of the Department of Constitutional Affairs

House of Lords	• continuing reform
	• reformed role for the Lord Chancellor
Judges	• setting up an independent selection body for judges
Courts	• creating a Supreme Court
	• restructuring of the courts system
	• specialist courts to deal with anti-social behaviour now stands at 41 (previously 12)

Department for Culture, Media and Sport

The Department for Culture, Media and Sport has a mixed portfolio ranging from ancient monuments to museums to grant aid.

Box 32.4 Main functions of the Department for Culture, Media and Sport

Historic structures	• protection of buildings by listing
	• scheduling of monuments to protect them
	• grants for the above
Funds for culture, heritage and sport	• national lottery policy
	• liaise with relevant funding bodies
	• listed building grants

Department for the Environment, Food and Rural Affairs

Box 32.5 Main functions for Department for the Environment, Food and Rural Affairs

Environment	• coastal management policy, including coastline defence works
	• contaminated land and the remediation regime
	• waste management policy
Agriculture	• application of European agricultural policy
Rural affairs	• problems of affordable housing in rural areas
	• the ban on the traditional hunting of foxes
Fisheries	• problems of over-fishing
	• the decline of the fishing industry

Department of Trade and Industry

Box 32.6 Main functions of the Department of Trade and Industry

EU regional aid	• development of strategy for period from 2007 to 2013
Trade	• policy for the World Trade Organisation
	• policy for the EU
Industry	• policy for small and medium enterprises
Highways	• highways agency
Natural resources	• oil and gas exploration and development landward and North Sea
Energy	• renewable energy policy
	• wind farm developments — landward and at sea

Agencies

Government departments are the major components responsible for the delivery of government functions. In the main, the work is undertaken by the civil service hierarchy within the department. However, departments sponsor agencies, *ad hoc* committees, task forces, public corporations and other bodies. They may do this solely but sometimes severally with another department or even outside bodies. The agencies need to be considered in that they exist to support the work of a department and many have immediate relevance to the property industry.

Many government operations are carried out by agencies rather than departments, the former being linked to the latter in a number of ways. Various terms are used to distinguish the agencies, such as

---

Box 32.7 Typical government agencies and non-government government bodies involved with the property industry

English Heritage		• heritage estate
		• statutory consultee
		• listed buildings operations (subject to legislation)
English Nature		• nature conservation
Highways Agency	Executive Agency (Transport)	• strategy for motorways and trunk roads
Valuation Office Agency	Inland Revenue (Treasury)	• national taxation (see Chapter 27)
		• local taxation (see Chapters 28 and 29)
		• valuation for taxes (see Chapter 30)
Regional Development Offices	ODPM	• regional economic and social development
		• regional planning and development
Urban Development Corporations	ODPM	• economic and social regeneration of their area

---

advisory, executive and independent. Box 32.7 identifies some of the many main property related agencies, examines them and refers to other chapters where they are considered in context.

Thus, the agencies are distinguished by their function as follows:

- executive agencies — they carryout executive functions as a separate entity within a department, having their own departmental staff and a budget
- executive non-departmental public bodies — they are not within a department, but in carrying out executive government functions at arm's length from the minister, they are of an administrative and regulatory nature (with their own outside staff and budget)
- advisory agencies are created to advise a minister, generally not having their own staff — being supported administratively by the staff of the department.

Future

Buildings of historic or architectural interest

It seems that English Heritage will take on the day-to-day operations for listed buildings and for buildings with potential for listing — so being afforded protection from harmful works or demolition. The Secretary of State for Culture, Media and Sport will retain overall policy control. The proposed change may require legislation.

Government Funding

Aim

To describe the way in which government funding relates to the property industry

Objectives

- **to describe the nature of government fund raising**
- **to explain the reasoning for public, private and voluntary provision**
- **to demonstrate how funds are distributed**
- **to show how funds are generated for settlement and community development**
- **to identify and review future changes to the funding**

Introduction

Much of the money raised by central government is allocated to:

- government departments, agencies and other non-government public bodies
- regional, county, district and other local government authorities.

Issues on the allocation of centrally held funds to the regions, counties and other tiers of local government hinge on such matters as:

- the appropriateness of the formulae used to allocate funds to local populations
- accountability for funds which are not derived locally
- the burden on local government for items considered as national.

Although these matters are not directly relevant to the property industry, they may have implications for it, particularly as locally raised funds are usually a tax on business and residential property occupation. In the case of business property the amount raised in rates is £19 billion; council tax on dwellings raises £20 billion.

Raising funds for government

Fund raising by the government comes from several sources, such as:

- taxation and national insurance receipts
- borrowing by the issue of government gilt edge stock
- revenue surpluses from property, business or commercial operations, eg rents, fees or commissions and sales
- disposals from privatisation of statutory business holdings, eg government's "golden" shareholdings
- capital receipts for the sale of land and other surplus assets.

Estimates of government expenditure and funding

Broadly, the Chancellor of the Exchequer determines general level of government expenditure in the light of on-going and forthcoming approved departmental programmes. Funding for the work needs opportunities and constraints to be taken into account, including:

- the general state of the economy
- the prospects for growth, inflation, employment, balance of trade and the like
- the price of oil
- estimates of the above-mentioned receipts from sales of assets and so on
- the estimated level of taxation receipts.

Some £455 billion are collected by the national and local taxation authorities. In effect, the taxes and other funds (other than taxes and other monies raised by the Scottish Parliament and the Welsh Assembly) are allocated to government departments for redistribution to:

- government departmental budgets — covering executive agencies and other internal departmental projects and programmes
- the budgets of executive, advisory and other non-governmental public bodies — in fact, the funds are distributed out of the departmental allocations
- the budgets of the Scottish Parliament and the Welsh Assembly
- local government budgets — this includes the non-domestic rates (business rates) and a slice of the national tax receipts.

The actual amounts from the principal national and local taxes are shown in Box 31.5.

Council tax receipts and business rates

About £20 billion of council tax is collected by the local government charging authorities and certain other bodies for direct expenditure by the charging and precepting bodies. Generally, the local bodies are free to fix the level of taxation but there is a ministerial power to cap the amount any particular authority intends to charge. (Additional funds, amounting to about £19 billion, come from the business rates which the local billing bodies had remitted to the government — the amounts are redistributed on a capita basis according to prescribed formulae.)

Box 33.1 Amounts of tax from the national and local sources of taxation

	Tax	£ (billion)	£ (billion)
National taxes	• excise duties	40	
	• corporation tax	35	
	• income tax	128	
	• national insurance	78	
	• VAT	73	
	• other taxes	62	416
Local taxes	• council tax	20	
	• non-domestic rates	19	39

Audit and CPA

Under the Audit Commission Act 1999, in addition to its duties to run CPA (measuring the performance of public bodies), the Audit Commission has powers to audit their accounts.

Borrowing

Any shortfall from the estimates of receipts and expenditures will give a figure for estimated borrowing. Actual borrowing by the issue of stock will arise on occasions during the year, as necessary.

Issues of funding

Tax base

An area of particular interest and concern to those in the property industry is the tax base. In recent years it has been developed with many new taxes. It seems that society finds it difficult to settle on the tax base for raising funds for goods and services provided centrally and locally. For at least 400 years, the principal source of locally raised funding had been rates; largely based on the rate poundage multiplied by the net rental value of property. Modern day national taxes began about 200 years later with income tax to fund the Napoleonic War. Current sources of taxation are:

- national income and corporation taxes
- national taxes, eg capital gains tax on gains and inheritance tax on the estates of deceased individuals
- various environmental taxes
- value added tax and customs and excise duties
- stamp duty and stamp duty land taxes
- rates payable by businesses (national non-domestic rate multiplier times the net rental value of the businesses property) and occupiers of dwellings in Northern Ireland

- council tax payable by occupiers of residential accommodation (a tax charged according to the capital value band in which the property was designated)
- finally, national insurance contributions.

Almost every year the government, sometimes as a result of EU policy, finds other ways enhancing the tax base!

Public, private or other funding

An important funding issue for society is the decision on whether a necessary common service or infrastructure should be in the public, private or voluntary sector.

Until the 1980s the public sector was the principal domain of infrastructure and certain other businesses, eg forestry, and systems. From that time privatisation has become an established policy so that much as been transferred to the private sector. In some instances this has been to property investment companies, eg by sale and leaseback, or to contractors by private finance initiatives.

For several centuries, central government has always funded and controlled the military and this is to continue. During the 19th century a mix of public local authority and private sector provision became established for utilities, but during the 20th century the many small organisations offering services abruptly or slowly changed into large public corporations, including:

- British Railways Board
- National Health Service (NHS)
- British Overseas Airways Corporation
- Post Office
- Central Electricity Generating Board.

Privatisation

However, since the 1980s various devices have been used so that a proportion of the public sector has been privatised. The approaches used have included:

- de-nationalisation, eg airports, air passenger services, electricity (generation, transmission, and sales), gas, railways; telecommunications and water
- outsourcing of public services to the private sector, eg management of prisons, management of public buildings
- disposals of both surplus and operational landed estates, eg Forestry Commission sales of commercial woodlands, the new towns was effected to the private sector or to local authorities
- private finance initiatives to fund major projects with various ownership and management configurations, eg hospital buildings in the private sector and management in the NHS
- outsourcing government services to the private sector
- the creation of self- or largely self-financing government agencies, eg the Land Registry, Her Majesty's Stationery Office and Ordnance Survey.

Property sales

Land and property sales have been going on for a very long time. Examples abound but a few are:

- sales of surplus non-operational land by British Railways Property Board and its successor
- sales of commercial and other woodlands by the Forestry Commission
- sales of surplus land from the NHS estate
- disposal of the new towns by the former Commission for the New Towns.

Broadly, this has meant that the capital in the buildings and businesses of much of the old public sector has gradually shifted to the private sector.

Private finance initiative

More recently, many, if not all departments, have become involved in the general ethos of the private finance initiative (PFI) in two ways, namely:

- developing the policy structure for PFI
- adopting a management or overseeing role, *vis-à-vis* the local authorities and other organisations within their remit that are adopting such projects
- in developing projects for internal purposes.

The Chief Secretary to the Treasury has a role in the area of developing policy and dealing with the fiscal context for PFI.

Voluntary sector

At the same time the provision of housing, parks, schools and colleges remained with local authorities. However, less demonstrably, a similar quasi-privatisation has been going on in this sub-sector. Thus, the local authority capital in housing has largely shifted to the registered social landlords who are funded as follows:

- grants or loans by the Housing Corporation
- loans from the financial institutions
- tenants' rents and charges.

Finally, many services which might fall to government are provided by organisations in the voluntary sector. Many of these organisations have charitable status and, therefore, enjoy a favourable tax regime — in effect public funds. Nevertheless, a substantial contribution is made to health, welfare, education and other causes by charities and other not-for-profit organisations.

Also, some local authorities are shifting leisure facilities to trusts for operational and financial reasons. Finally, projects which would probably have been in the local authority domain are now in the voluntary sector under a variety of partnership funding arrangements — much of the funding having come, and still coming, from the National Lottery and other funding bodies. Here, the National Lottery funding bodies may now be seen to be going into areas traditionally seen as pure public sector.

Funding of infrastructure and buildings

Funding of private sector accommodation and structures of all types is dealt with in Chapter 25. This chapter highlights public sector funding of infrastructure projects and buildings in the public sector and the voluntary sector's property which is partly public funded, eg housing through the Housing Corporation.

Thus, the government disburses funds for national purposes from the Treasury to government departments. They allocate them to programmes and projects for the department's direct operations or through the executive agencies (departmental or non-government public bodies). Similarly, the national non-domestic rate is collected by the rating authorities and, in effect, passed to the government from where it is remitted to the hierarchy of local government bodies according to government formulae of local need. Some funds are then used for property projects.

At a simplistic level it could be said that taxation is raised by the government for both revenue and capital expenditure on public sector programmes and projects. Taxation revenues are of the order of £455 billion, about 36% of gross domestic product and projected to rise to 38% over the next two years.

There are many sources of funding for infrastructure and buildings in the public and voluntary sectors, including:

- housing — government funds from the Housing Corporation
- culture and sport — grants or funds from partnerships, eg the Arts Council, Sports Council, the National Lottery funding bodies and local authorities
- education, welfare and similar projects — charitable foundations (partly supported by tax exemption and reliefs)
- public utilities — fees and charges from customers, government grants and debt financing.

Generally, infrastructure and buildings in the assisted areas may be supported by funds may be available from the structural funds of the EU.

European Funds

EU funds

The current round of EU funding lasts for the period 2000 to 2006 and discussions are already looking at the strategy for the next period (see below "Future"). A part of the UK's value added tax fund is remitted to the European Commission where it is used for EU purposes, including the structural funds; some of which benefit deprived areas of the UK.

Structural funds

Part of the EU funding is divided and allocated to structural funds, namely:

- European Agriculture Guidance and Guarantee Fund (EAGGF)
- European Regional Development Fund (ERDF)
- European Social Fund
- Financial Instrument for Fisheries Guidance.

Objectives

Each fund allocates its allocated monies to geographic areas, regions or rural areas under objectives, numbered 1, 2, 3, 4, 5a, 5a, and 6. Box 33.1 identifies the objectives.

Box 33.2 Objectives under the EU structural funding policies

Objective	Characteristics of target areas
1	• regions with lagging development
2	• regions (industrial decline — serious))
3	• any area (unemployment — long-term)
	• any area (excluded groups — integration)
4	• any area with industrial change (unemployment)
5a	• agriculture (structural adaptation)
	• fisheries (structural adaptation)
5b	• rural areas (vulnerable)
6	• regions (population — extremely low density)

Hitherto, EU criteria for investment in a project depended on its location. Projects in, for example, tier 1 areas, such as Cornwall and West Wales, received 35% of the investment. The scale varied according to the severity of the regional problems in an area and could have been as high as 40%. Tier 2 areas received government incentives of up to 20%.

State aid for commercial projects

State aid for projects is severely prescribed by the EU so that that given supports EU policies for regional development and social development. International mobile industry is aggressively fought for by government agencies at national, regional and local levels. Nevertheless, state must not breach EU competition law.

Local funding

About 20% of local government funds are generated from the council tax and other local government sources. The other 80% is derived from central government. Of course, many government departments and agencies disperse revenues and capital funds into the local economy, eg the Housing Corporation's capital grants go to local social registered landlords for housing projects. This section is not comprehensive but is intended to give a flavour of the diversity of approaches to local funding. Box 33.3 shows a limited but novel range of funding initiatives.

Box 33.3 Examples of local funding opportunities

Planning gain or obligation	S 106 of the Town and Country Planning Act 1990 provides for an agreement to be made between a developer and the local planning authority. Such agreements must pass the five criteria given by the ODPM. They are often used by the local planning authority to obtain some local benefit for the community or a contribution to local funds.

Thus, s 106 agreements are frequently used to obtain funds, or works in lieu, from developers wanting to obtain planning permission. Terms used to describe the approaches used included planning gain, planning obligation and, community benefit.

Generally, a planning obligation, etc could not be made a condition of planning permission but could be enforced as a contract. Under such an agreement the developer promises to do certain things, such as:
- improve a highway access to the development site
- reserve land for a primary school on a site for housing development
- plant a screen of trees in a park abutting a housing development
- allocate some houses as affordable homes on a residential development site.

Planning contribution	The Planning and Compulsory Purchase Act 2004 introduced a new form of planning payment, known as the planning contribution. The local planning authorities are required to make up a schedule of rates for payments by developers. Subject to approval of the schemes by the ODPM, developers will be charged at £x per m² floor area for offices and other rates for other kinds of development unless they choose to negotiate a planning obligation.
Business Improvement Districts	BIDs are a means of implementing the planning function. The Local Government Act 2003 provides for a BID levy fund to be raised from business ratepayers with property in the BID. Others may make contributions to the fund, eg local authorities and other partners to the proposal.
Private finance initiatives	Numerous PFI schemes have been created by public sector bodies and private sector partners to invest private funds in projects required by the public sector. Buildings and operational projects have been developed in this way.
Independent living funds	The ILFs are discretionary trusts funded by the government. They fund support to enable individuals to live in their own homes. The individuals have disabilities which require a high level of personal assistance.
Pension trusts	Local authorities establish pension trusts for staff by using council tax receipts and any other contributions to acquire investments according to statutory prescriptions and the trustees' investment policies. Professional independent advice is obtained from property professionals and others.
National lottery	The various funds set up under the auspices of the National Lottery provided about £1.3 billion for a widened range projects (and £0.6 billion in taxes). In 2003–2004, the projects covered were: charities (£298 million), arts (£141 million), heritage (£328 million), sports (£104 million), Millenium (£1.1 million) and education health and environment (the new category) (£679 million).

Future

EU expansion

Generally, the 2004 expansion of membership of the EU is likely to result in an increase of the UK's contribution and, possibly, a loss or reduction in the special rebate to the UK. Also, as the average GDP has fallen a need will arise to adjust the special areas which are eligible for EU support.

Strategy for the next round

The ODPM is developing the national strategy for the way in which the next round of EU funds for projects and proposals will be strategically implemented across the EU. The period under review is from 2007 to 2012. Hitherto, assisted areas have been relatively large parts of the UK, such as Northern Ireland and North East Scotland.

There is an argument for refining the spatial distribution down to individual or cluster of abutting electoral wards for the UK or the equivalent area elsewhere. Such an approach will enable EU aid to be more precisely targeted throughout an EU member state.

Local taxation

Various official studies into local taxation have been made and at the time of writing (November 2004). Sir Michael Lyons is reviewing the funding of local government in England.

Generally, alternatives to property rates have:

- if adopted, been abandoned, eg community charge
- if mooted, not adopted, eg local income tax or local sales tax.

It seems that the current enquiry has two possibilities, namely:

- widen the tax base with other forms of taxation
- extend the bands of council tax.

However, for the moment business rates and council tax apply in Great Britain and rates apply in Northern Ireland. Following revaluation in Northern Ireland, it seems likely that changes will be made to the rating system applied to dwellings, eg capital values will be used.

In Scotland, the Scottish Parliament is considering a Bill to abolish council tax and replace it with a service tax (an income tax).

Part 10
Settling Disputes

Settling Disputes

34

Aim

To show how disputes involving property are resolved

Objectives

- **to describe the various roles and activities in dispute resolution**
- **to identify the approaches to dispute resolution**
- **to describe the main approaches to dispute resolution**
- **to analyse the factors determining which method is adopted for a particular dispute**
- **to outline the general conduct of a selected approach**
- **to examine the various courts, tribunals and other methods**

Introduction

Resolving a dispute follows a broadly similar pattern no matter what method is used. However, the different methods vary greatly in terms of the following:

- the speed with which a matter in contention is settled
- the degree of formality in the proceedings
- the number of advisers involved in the procedure
- the risks involved, particularly costs
- the prospect of appeal and hence delay in reaching finality.

This chapter looks at horses for courses — essentially which method or methods of dispute resolution would be appropriate for a particular property problem.

Disputes arise in the property industry in most situations. When contemplating a prospective or draft contract (if any) for an activity (window cleaning), project (construction of a building), or procedure (insurance), the kinds and nature of disputes which commonly arise need consideration.

In each instance, an assessment of the risk of dispute is required and the parties need to agree how any possible dispute should be resolved. Of course, such an approach requires an understanding of the approaches to dispute resolution and the principles which may be used to choose an appropriate method of determination; assuming that choice is available.

Roles and activities

Box 34.1 Roles and activities in dispute resolution

Claimant	• aggrieved person seeking a remedy to complaint
Defendant	• person complained about of alleged mis-doing
Judge	• a solicitor or barrister of requisite experience • appointed to be a judge by the appropriate authority • the court appoints a particular judge to hear a case • hears the evidence, reads the documents and hears the submissions of the parties or their lawyers • finds the facts and applies the law to reach the decision
Arbitrator	• a lawyer or specialist professional in the field to be heard • appointed by the parties, a professional body's president • hears the case • determines judicially • not liable as arbitrator, unless acted in bad faith or resigned (when liability may arise)
Mediator	• lawyer or other person trained in mediation process and procedure • acts impartially and does not get involved emotionally • initiates the day with an open meeting of all the parties • consults to establish the facts and clarify the issues for the parties (in plenary sessions of or separate sessions with the parties) • asks questions and points to the decisions which need to be made by the parties • all done to elicit progress to the resolution
Ombudsman	• appointed under statute or by an industry as knowledgeable of or experienced in the field • acts informally by correspondence • does not charge the complainant • establishes the whether complaint justified • award remedy, eg compensation
Adjudicator/independent expert	• probably a professional in the field of the dispute • appointed by the parties, perhaps under statute • hears evidence from the parties and decides — based upon his knowledge and experience
Expert witness	• professional with knowledge and experience • must do the best possible for truth • may use books, specialist material and so on • gives opinion — has a duty to the court
Witness	• person with knowledge of facts and the parties, etc

---

<table>
<tr><td>Box 34.1 continued</td><td></td></tr>
<tr><td>Solicitor</td><td><ul><li>lawyer who handles the case for one of the parties</li><li>any barrister is instructed by the solicitor</li><li>may be an advocate solicitor able to appear in court</li></ul></td></tr>
<tr><td>Barrister
(see solicitor)</td><td><ul><li>lawyer who appears in court for one of the parties</li><li>presents the case, calling the witnesses</li><li>cross examines other party's witnesses</li><li>sums up on behalf of client</li></ul></td></tr>
</table>

Provision for settlement of disputes

There are many arrangements for settling of disputes. A well drafted contract will cover the matter and statutes offer many more. Professional bodies and trade associations usually require their members to have mechanisms for dealing with complaints or disputes. Indeed, professional bodies and others make such arrangements, eg PACT for settling rent review disputes between landlords and tenants, which was set up by the Law Society and the RICS.

Some disputes arise from situations where there is no contract as such. Thus, some disputes, such as those concerning squatters, marriage breakdown or compulsory purchase, arise from non-contractual rights being hurt or abused. Sometimes a statute determines the approach to be adopted, eg on compensation for compulsory purchase, the Lands Tribunal. Otherwise, an after-the-event choice will need to be made.

Nature of cases

Professional or building work

Disputes about contracts for professional or building work are at least two levels of concern:

- first, concerns involving alleged inadequate conduct giving rise to a complaint
- second, concerns involving an allegation of breach of contract (which may be negligence as well).

By definition complaints are dealt with out of court, usually by the sole practitioner or by the company or firm causing the concern (see below). The second type of concern is likely to be about a more serious matter and one which frequently requires the advice of a solicitor should the other party not admit liability and settle the matter forthwith.

Other disputed matters

Other disputed matters arise as concerns about other relationships involving a vast range of matters — almost everything could be mentioned. Instances include the following:

- landlord and tenant relationships, eg rent reviews, enfranchisement price, compensation for tenant's improvements
- compulsory purchase, eg compensation for disturbance
- marriage breakdown, eg division of the marital property
- insurance claim, eg eligibility to reinstatement
- rights of way, eg an inadvertently blocked footpath.

Box 34.2 briefly portrays a selection of torts arising in the property industry which, unless the parties settle the dispute themselves, may need to be addressed in some form of dispute resolution. The treatment in the box is necessarily brief; a tort should be addressed by each of the parties with the appropriate level of advice and support from a qualified and experienced legal practitioner. Whereas claimants have allegedly suffered loss, damage or injury as a result of the defendants' actions, the defendants may find comfort that there are defences against the allegations which may remove or mitigate the claim. Finally, the box does not cover criminal matters or, indeed, every kind of civil dispute — only selected torts.

Criminal matters

The perspective of this volume is disputes of a civil nature, ie none of these matters are likely to result in criminal proceedings. Criminal matters are for the police and the criminal courts. There are some areas where those in the property industry have a duty under the law to handle appropriately suspected criminal matters, including:

- the prevention of insider dealing, eg creating Chinese walls within a professional practice
- acts of property misdescription by estate agents or developers
- some statutory offences which may be civil matters but also have a criminal aspect or penalty
- money laundering, eg estate agents (and others) are required to set up compliance procedures to report suspected instances
- illegal use of premises, eg on compulsory purchase the illegal use of premises must be ignored in assessing compensation.

Approaches to dispute resolution

Box 34.3 gives the approaches commonly available in the property industry and includes brief description of each.

The parties in dispute may find that the choice of approach is determined by one of several means, eg the law, the contract between the parties. It is important, therefore, that the parties to any proposed relationship involving property should consider, among other things, the kinds of, or the ways in which disputes may arise. Approaches which could be used to resolve any dispute may then be more readily selected. The selection factors are given in Box 34.4, but Box 34.3 is pertinent.

Box 34.2 Alleged torts and other dispute generating events concerning property

Tort of negligence — duty of care	• professional consultant	• duty of care exists • causes damage by act or omission
	• contractor	• there is proximity between the parties • damage must be foreseeable
	• sub-contractor	• remedying is just and not unreasonable (against the test of an ordinary skilled person exercising and professing to have a special skill)
Vicarious liability	• employee	• employee commits tort • does so in the course of employment
	• employer	• liable to claimant for employee's tort • employer probably has insurance
Tort from land — occupier's liability	• authorised visitors	• the Occupiers' Liability Act 1957 seeks to protect authorised visitors • imposes a duty of care • children given special care
	• trespassers	• the Occupiers' Liability Act 1984 seeks limited protection for trespassers • protection is limited to personal injury (from premises)
Tort from land — nuisance	• private nuisance	• use of own land interferes unreasonably with private use or enjoyment of another's land, etc
	• public nuisance	• large class of citizens affected • loss or damage suffered by the claimant
	• statutory nuisance	• statute provides for the nuisance • actionable by local authorities • non-compliance cases heard by magistrates
Tort on land — trespass	• trespasser	• trespass actionable • trespasser causes damage • damages actionable
Tort of strict liability	• animals under common law	• known dangerous animals and escaped animals may cause damage • liability arises at common law
	• animals and statutes	• Animals Act 1971 covers much of the common law • dangerous and non-dangerous species covered • defences against liability include claimant accepted the risk

---

Box 34.3 Approaches to the resolution of disputes

Litigation	• a formal procedural approach of some gravitas in a court setting
	• many parties are represented but a litigant may find the other party is not represented
	• unrepresented litigants may add to the complications of the proceedings and the expense involved
	• loser may be required to bear the costs of both parties
	• judge and advocates are lawyers by profession
	• appeal to higher courts
Lands Tribunal	• less formal approach but somewhat legalistic
	• non-lawyer professional may determine a case or a lawyer with technical colleagues
	• appeal to the Court of Appeal on a point of law
Arbitration	• determination likely to be by a specialist professional trained in arbitration
	• acts in a quasi-judicial manner, ie not using own knowledge and experience but settles on the basis of the evidence presented
Expert determination	• specialist professional may receive evidence
	• settles on the basis of own knowledge and experience
	• may be sued for negligence
Mediation	• mediator acts as a go-between for the parties
	• offers guidance approach and matters which needs to be addressed
	• seeks to get the parties to an agreement
	• parties may drop-out of the proceedings at any time
Complaints procedure	• takes complaints in a stepped approach within the branch or firm and then to independent determination
	• does not cover negligence, fraud and similar issues
Small claims court	• a very inexpensive approach
	• cases involving a prescribed sum
	• parties not often represented
	• no appeal to a higher court
Ombudsman	• industry-based determination of complaints not settled by the firm or company
	• relatively informal
	• power to award compensation
	• complainant may take case to court if not satisfied.

Generally, the risks inherent in any activity, project or procedure should be examined and dealt in ways which should avoid disputes. But that is an ideal world.

In this chapter the principles for choosing a method of dispute resolution are dealt with in the first instance and this will be followed with an appraisal of the methods listed above

Principles

The principles listed above are dealt with in this section. However, there is a general presumption that the parties will want to avoid litigation, principally on the grounds of cost.

Box 34.4 Factors affecting the choice of the approach to resolving a dispute

Nature of the dispute	• the law provides the approach to settling some disputes, eg the Lands Tribunal for compensation for certain planning restrictions • the contract between the parties may do likewise, eg a lease may provide for arbitration or for independent expert as an alternative
Fees and costs	• courts may impose all or some costs against the person who loses the case • Civil Procedure Rules allow the court to penalise litigants who are obstructive
Speed of determination	• the speed of settlement may be essential, eg an issue on an on-going building contract
Insight and rigour of the determination	• rigorous legalistic approach tends to be taken by the courts • ombudsman's approach tends to be on a fair and reasonable basis • mediation is an approach where an expert guides the parties towards a settlement
Applicable law and jurisdiction	• the applicable choice of law and jurisdiction given in the contract is an important point to settle, probably by agreement (particularly with overseas contractors or suppliers)
Degree of formality	• the range is high formality in the courts and relatively low formality in ombudsman's work • mediation, eg in the division of family assets on divorce is very informal
Enforcement	• judgments of the courts may be enforced by execution against assets • contempt of court proceedings may be applied where a party acts against an order of the court • otherwise varying with the approach
Appeal.	• appeal available in the courts (subject to permission, etc) • appeal from an arbitrator award is limited, usually to procedural misconduct or an error of law • from an award of the Lands Tribunal to the Court of Appeal on a point of law

The nature of the dispute

In some instances a court hearing or litigation cannot be avoided, particularly where criminal activities are alleged. In other areas of law, statute lays down that a court or tribunal should determine a matter.

Where a dispute involves criminal activities the matter will be settled, perhaps in part, under the criminal justice system and will be heard before a judge in court. A large number of disputed (non-criminal) property matters will be heard by a court or tribunal either as a result of statute or as provided under a contract (which is silent as to other methods of dispute resolution) between the parties in dispute, as indicated in Box 34.2, which shows something of the range of such matters. Of course, despite a requirement to go to court under their contract, the disputing parties may be able to agree to adopt an alternative method of resolution.

Independence and appointment

It is essential that each of the parties perceives the person who will resolve the dispute is independent of them and that he has no personal interest in the outcome. Independence is taken for granted in litigation; the administration of the court or tribunal appoints the judge or the member respectively.

For other forms of dispute resolution, this is usually achieved by the parties either jointly agreeing a mutually acceptable individual or agree to an independent organisation, eg the Law Society or the RICS, making the appointment.

Costs

Three kinds of costs arise, namely:

- the costs of professional representation, if any, eg lawyer or surveyor
- the costs of any witnesses
- the fees and costs of the person determining the dispute.

Generally, the cost of litigation is relatively high, sometimes extremely so. If possible, litigation should, therefore, be avoided by agreeing the dispute without recourse to an outside party.

In the event, however, each party may bear their own costs and share the costs of the hearing but more usually the question of costs is determined by the court or tribunal. Sometimes legal aid may be available, and at other times the dominant party may agree to pay the costs of the other party.

In other instances, the burden of costs may be settled in one of the following ways:

- the person appointed to determine the dispute apportions the costs
- each party agrees to bear their own costs and share the cost of the determination
- costs are shared equally
- one party agrees to bear all the costs.

In the absence of legal aid, the prospective of costs probably deters some citizens from seeking redress by way of dispute resolution, particularly where the prospect of success is not certain.

No-win no-fee

Some lawyers will proceed with a case involving a claim for compensation on a contingency fee basis, ie unless the outcome is successful no fee will be paid. The level of fees in cases where the claim is successful can be high, possibly of the order of 30% of the compensation awarded.

Urgency of resolution

The context of the dispute may raise urgency as a factor in determining which approach is used to settle a dispute. If the context is an on-going construction job which would have to stop until the dispute on, say materials used, is settled, the resolution must be achieved quickly.

Degree of formality

The parties may want a relatively informal approach to the settlement of their dispute. Generally, court hearings are more formal than tribunal proceedings, which are in turn relatively more formal than other methods of determination.

Remedies

For justice, the injured party, the claimant, should have a proven claim remedied against the other party or parties, the defendant(s). The kinds of remedy must suit the circumstances — a number of remedies are available to the courts or other resolvers of disputes, such as:

- stopping the defendant doing something
- requiring the defendant to restore the claimant's position
- requiring the defendant to pay compensation
- requiring the defendant to pay the claimant's legal costs and fees.

Enforcement

The enforcement of a court or tribunal decision is often a matter that the court or tribunal will act upon where contempt of court is alleged. Otherwise the party seeking enforcement will have to seek a court order.

Appeals

A court or tribunal decision may result in an appeal (and hence delay). In some instances, a court may state that an appeal is not allowed or may require that leave of court or tribunal.

Other methods of resolution may result in a final decision without appeal. (In a case where a party alleges negligence on the part of the individual determining the dispute, the remedy may be to take action for damages.)

Conduct of civil court cases

Generally, in the first instance, civil court cases concerning property will be heard by:

- a district court — the majority of civil cases are heard by district judges
- the county court, or
- the high court (including the Lands Tribunal).

Appeals go to the Court of Appeal and hence the House of Lords. In most cases the Civil Procedure Rules 1998 will apply.

In rare circumstances, the case may warrant appeal to the European Court of Justice, on EU matters or the European Court of Human Rights under the European Convention on Human Rights. Although few cases go to these courts, it is not an infrequent happening that human rights are raised in property matters in the courts here.

Civil Procedure Rules (CPR)

The CPR enables the court to govern the conduct of a civil court case before it with the general aim to deal with the case speedily, properly and at a reasonable cost. The court has power to charge costs and will penalise any unbecoming conduct of a party. The court wants to decide justly and therefore, under the CPR requires the parties cooperate to:

- identify the issues which will be presented
- gather and share relevant information
- sort out the issues upon what the court is to decide
- act speedily during the case.

The essence of the approach is shown schematically in Box 34.5.

Box 34.5 Schematic description of a court case under CPR

Step	Role	Activities
1	claimant	• prepares claim and, if necessary particulars of claim • adds a statement of truth • sends them to the court with copy or copies for the defendant(s) • sends fee to the court
2	court	• gives the case a claim number and prepares a file • sends documents to the defendant
3	defendant(s)	• responds with acknowledgement • responds with defence against the allegation • responds with an admission
4	court	• hears the submissions
5	court	• decides the case
6	parties	• consider the judgment • settle or decide whether to appeal

---

Arbitration

Arbitration is along established means of dispute resolution with its own statute-based law and practice. The Arbitration Act 1996 is now the principal statute on disputes arising from an arbitration agreement, ie

> an agreement to submit to arbitration present or future disputes (whether they are contractual or not)

Much arbitration is generated under contract but some statutes provide for them.

In the property industry arbitration is a common form of dispute resolution; frequently incorporated into various kinds of contract. Thus, a clause for the adoption of consensual arbitrations is commonly found in:

- leases
- building contracts
- contracts for the supply equipment for building services, oil and gas installations and the like.

The principal professional bodies hold lists of members who arbitrate on specialist areas. A solicitor drafting a contract will most likely insert a draft standard clause when drawing up, say, a lease for the president of a specified professional body to nominate a person to be the arbitrator should a dispute arise and the parties call upon the president to do so. The membership of the Chartered Institute of Arbitrators has specialists in a variety of property industry related topics who may be so nominated. Also, the Institute runs independent arbitration services for some professions or business areas.

Hearing

A lone arbitrator (or two arbitrators) take evidence from the parties to the dispute and determines the matter in a judicial manner; in the sense that only the evidence produced may be used, not an arbitrator's own knowledge and experience — any arbitrator is almost certain to be a professional in the field in question.

Adjudication and expert determination

Expert determination is a method of dispute resolution where an independent expert is nominated by the parties or by an independent organisation. Essentially, the expert will receive evidence, make personal investigations and use person knowledge and experience to reach a decision. This approach is relatively informal and inexpensive.

Adjudication is a similar approach which applies in many construction disputes, eg by virtue of the Housing Grants, Construction and Regeneration Act 1996.

Valuation tribunal

The 56 valuation tribunals, replacing the valuation and community charge tribunals, are local bodies comprising three members drawn from a panel of local people. They hear appeals against the Valuation Office Agency's local taxation assessments of property. The valuation tribunals cover both business rates and council tax.

Where the valuation tribunal accepts the appeal it may direct the valuation office agency to change the assessment and alter the list; otherwise it dismisses the appeal. A party to a hearing may appeal against the valuation tribunal's determination by going to the High Court or Lands Tribunal on points of law as appropriate and rating valuation matters to the latter.

Administration and hearings

For a one-off appeal the clerk to a valuation tribunal arranges hearing and during the sitting will advise the members on procedures and points of law, if necessary. Hearings are conducted in a relatively informal manner, usually with the ratepayer or taxpayer and the Valuation Officer appearing in person.

Valuation tribunal service

The Valuation Tribunal Service was established under the Local Government Act 2003 as a non-departmental public body (NDPB). From 1 April 2003, it provides administrative services, procedural advice, training, etc to the valuation tribunals.

Lands Tribunal hearings
Jurisdiction

Valuation matters constitute the jurisdiction of the Lands Tribunal. Awards are made on points of law with appeal to the Court of Appeal but the Lands Tribunal is the final determiner of fact on valuation disputes. Most of the cases concern statutory valuations but the Lands Tribunal Act 1976 empowers the tribunal to arbitrate on matters referred to it.

Its jurisdiction includes valuations for:

* business rating (see Chapter 30)
* compensation in cases involving compulsory purchase (see the Land Compensation Act 1961, as amended)
* compensation due to loss due to the physical factors caused by the coming into use by a public project, eg noise, fumes, etc due to motorway traffic (see the Land Compensation Act 1973, Part I)
* compensation for adverse planning decisions, eg a discontinuance orders; modification orders and revocation orders (see the Planning and Compensation Act 1991)
* national taxation including capital gains tax and inheritance tax (see Chapter 30)
* statutory wayleaves and easements, eg pipelines under the Pipelines Act 1962
* awards on the discharge and modification of restrictive covenants, including compensation, if any (see s 84 of the Law of Property Act 1925).

Appeals

An appeal may be made on a point of law (not fact) from the Lands Tribunal to the Court of Appeal.

Mediation

Mediation is a form of guided settlement where the mediator acts as a neutral chairman and as a facilitator or go-between; often between parties who have become entangled in disputed facts and views in an emotional way. As a go-between, the mediator will move continuously from one party to the other, bearing information, views and enquiries (which he or she may have to present in a clearer and better structured manner than might have been the case).

The mediator will open the day with a meeting of the parties to explain the mediator's role and the way in which the sessions will be conducted. Bringing the parties together from time to time, facilitates guided face-to-face exchanges; again keeping the meetings in an orderly, flowing and targeted manner.

Key to the situation is impartial objectivity and not becoming emotionally involved in the parties. Experience and knowledge should enable each party's presentation to be given in an orderly and precise manner. This should draw out the salient facts and views, so enabling the parties to see the solution to their dispute.

The approach is relatively informal and inexpensive and is more likely to be acceptable in what may be trying and uncertain times for the individuals involved.

Complaints procedures

Many individual professionals companies and firms or professional bodies have complaints procedures which are designed to prevent full-bodied disputes arising. They should be conciliatory rather than adversarial. Their jurisprudence may be limited in that they will deal with such matters as professional misconduct rather than negligence — the latter being a matter for court proceedings.

Some professional bodies have set up procedures for dealing with complaints against their members in those instances where the member's first level complaints procedure has failed to settle the matter to the complainant's satisfaction. Normally, they will not deal with professional negligence.

As noted above, some bodies have organised an arbitration service with the Chartered Institute of Arbitrators.

Compensation schemes

Apart from the awards of compensation for loss made by the ombudsmen, the property industry has several direct or indirect compensation schemes linked to it. They include:

- the Investors' Compensation Scheme
- the Policyholders' Protection Board.

Ombudsmen

A number of statutory offices of ombudsmen are related directly to parts of the property industry. Other ombudsmen deal with industries or sectors of the economy which include or impinge on property. Thus, the ombudsmen associated with the property industry are:

- the Ombudsman for Estate Agents, to which an estate agent may join voluntarily
- Financial Services Ombudsman
- the Legal Services Ombudsman.

(As a result of the development of the Financial Services Authority, the Financial Services Ombudsman was created out of the ombudsmen for the banks, building societies, insurance industry and investment industry.)

Broadly, the principles under which each ombudsman works are common to each of their offices. Their services are free and informal. They usually conduct a complaint by correspondence, not requiring professional representation on behalf of the parties. It is an alternative to going to court and will not be available in cases where there has been a court hearing. Generally, any ombudsman's award is respected but is not necessarily enforceable by the complainant.

A typical step-by-step approach to taking a complaint to an ombudsman's office is shown in Box 34.6. Guidance on how to handle what is usually an informal process is available from each ombudsman.

Box 34.6 Schematic approach to taking a complaint to an ombudsman		
Step 1	Approach the company	• explain the problem and seek redress form staff • if not solved, ask for official complaints procedure • if not solved, contact the ombudsman within set period of final response
Step 2	Approach to the ombudsman	• company's response is not acceptable or it has not responded in set period • receive advice on how to proceed
Step 3	Ombudsman sends complaint form	• complete the complaint form and send it in with relevant documents
Step 3	Follow ombudsman's initial approach	• ombudsman may seek to mediate informally • if not resolved, follow ombudsman advice
Step 4	Ombudsman appoints adjudicator	• adjudicator seeks further information from parties and others in carrying out a full investigation • adjudicator determines the outcome and informs the parties
Step 5	Ombudsman's award received	• if adjudicator decides for the complainant, up to set limits may be awarded

Public services complaints

The public sector has three officers of ombudsmen to receive and determine complaints from members of the public, namely:

- the Local Government Ombudsman, in England
- the Ombudsman, in Wales
- the Scottish Legal Services Ombudsman.

Part 11

Appendices

Appendix 1

Table of Boxes

Appendix 2

Table of Statutes

Table of Statutory Instruments

Appendix 3

Table of Cases

Appendix 4

Table of Abbreviations

AIM	Alternative Investment Market
AME	annually managed expenditure
ARLA	Association of Residential Letting Agents
ASB	Accounting Standards Board
ATCM	Association of Town Centre Management

BCIS	Building Cost Information Service
BID	business improvement district
BMCIS	Building Maintenance Cost Information Service
BPF	British Property Federation

CCTV	closed circuit television
CDM	Construction Design and Management
CGT	capital gains tax
CIC	community interest company
CIOB	Chartered Institute of Building
CORGI	Council for Registered Gas Installers
CPA	comprehensive performance assessment
CPD	continuous professional developement
CPR	Civil Procedure Rules

DCMS	Department for Culture, Media and Sport
DEL	departmental expenditure limit
DEFRA	Department for the Environment, Food and Rural Affairs
DRC	Disability Rights Commission
DTI	Department of Trade and Industry

EAGGF	European Agriculture Guidance and Guarantee Fund
EIA	environmental impact assessment
ERDF	European Regional Development Fund
EU	European Union

EUV	existing use value
EZ	enterprise zone
FRI	full repairing and insuring
FRS	Financial Reporting Standards
GAAP	Generally Accepted Accounting Standards
GPDO	General Permitted Development Order
HCR	home condition report
HIP	home information pack
HM	Her Majesty's
IASB	International Accounting Standards Board
ICT	information and communications technology
IHT	inheritance tax
ILF	independent living fund
IPO	initial public offering
IR	Inland Revenue
JCT	Joint Contract Tribunal
LNG	liquified natural gas
LPA	local planning authority
LT	Lands Tribunal
MV	market value
NAEA	National Association of Estate Agents
NALS	National Approved Letting Service
NHBC	National House Builders Council
NIMBY	not-in-my-back-yard
NO	number
NRAC	National Register of Access Consultants
ODPM	Office of the Deputy Prime Minister
OEA	Ombudsman for Estate Agents
OMV	open market value
ONS	Office for National Statistics
OS	Ordnance Survey
PFI	private finance initiative
PFS	petrol filling station
PLI	public liability insurance
P-P	Practical Points
PP	planning permission

PPG	Planning Policy Guidance
PPS	Planning Policy Statement
QC	Queen's Counsel
R&A	Roles and Activities
REFS	Really Essential Financial Service
RIBA	Royal Institute of British Architects
RICS	Royal Institution of Chartered Surveyors
RTPI	Royal Town Planning Institute
RTM	right to manage
S	section
S-by-S	Step-by-step
SDLT	stamp duty land tax
SHIP	Safe Home Income Plans
SI	Statutory Instrument
SWMP	site waste management plan
SWOT	Strengths-Weaknesses-Opportunities-Threats
TEGoVA	
T&CP	Town and Country Planning
UCO	Use Classes Order
UDC	urban development corporation
UK	United Kingdom
VAT	value added tax
VO	valuation officer
VOA	Valuation Office Agency
WLAN	wireless local area network(ing)
WM P	waste management plan
YP	years purchase

Part 12

Indexes

Index 1

Places

Index 2

Organisations, Offices, Officials and Others

Index 3
Key Words